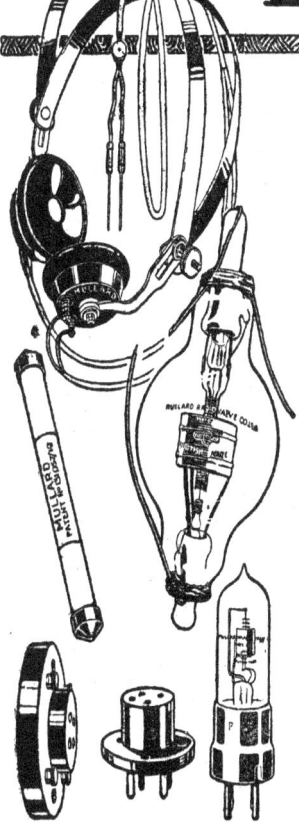

# Stories that Authors DARE NOT write

One Shilling net

True Story
Magazine

Truth is Stranger Than Fiction

The Editor offers—
£250
for your life Romance

Posed by Mary Miles Minter

A large illustrated monthly Magazine beautifully printed on high surface paper. For *the first time* it presents stories taken from the actual experiences of life. The writers of stories that bare the inner secrets of the human heart must necessarily remain anonymous, but their names will be scrupulously recorded and preserved as a warranty of their truth. The Editors expect the Magazine to be edited by its readers, who are therefore invited to send in their life stories, or any story they know to be true, which, if accepted, will be paid for at reasonable rates. SEE A COPY OF THIS MAGAZINE *TO-DAY.*

# *EVERYTHING*
## *for* **WIRELESS**

You can send to Ashley Radio with the confident knowledge that whatever you want for wireless will be supplied. Our stock is thoroughly complete from batteries to valves. Let us know your wants for wireless and we can supply them.

- Valves
- Crystals
- Ebonite
- Condensers
- Accumulators
- H.T. Batteries
- Induction Coils
- Filament Resistances
- Patent Coil Holders
- Etc. Etc.

It is BY FAR the largest and longest monthly Magazine offered for 7d. All the stories grip and are worth reading, and the Publishers believe that by adding this Magazine to those you read monthly you will derive more pleasure than has hitherto been possible for such a nominal outlay. Every number is filled with strong gripping fiction.

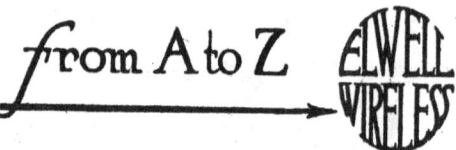

# from A to Z — ELWELL WIRELESS

## and the Guarantee

### Oxford and Cairo--
The first Imperial Wireless Chain Stations

### Northolt (nr. Harrow)--
For British G.P.O. traffic to Central Europe

### Horsea Island (Portsmouth)—
British Admiralty Station working to the Fleet all over the World

### Eiffel Tower—
Two 100 kw. Elwell Transmitters with record range of 11,810 miles to New Zealand

### Lyons—
Two 350 kw. Elwell Transmitters working regularly to U.S.A.

### Rome—
With three 714ft. wooden masts, the highest wooden structures in the world — works regularly to U.S.A.

and many others, totalling 60 large masts and 1,100 stations

### Wireless Broadcasting—
Elwell broadcasting receivers and Components have the same high efficiency and reliability as the above Elwell high power stations

We invite you to write for descriptive folder of our broadcasting equipment

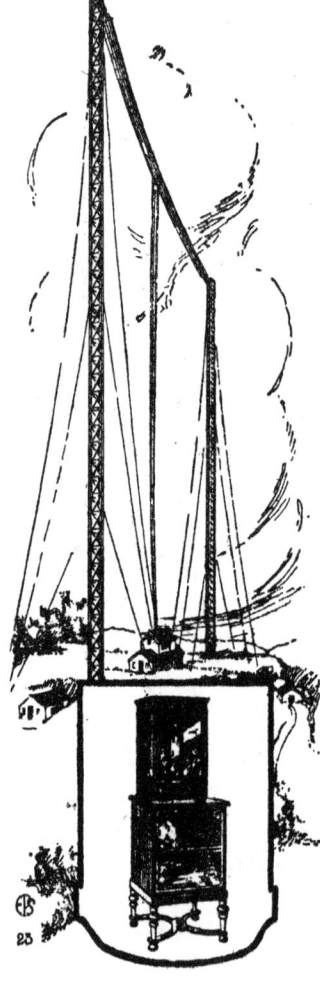

## C.F. ELWELL.
Craven House, Kingsway. London. W.C.

Telephone: Regent 1061
Regent Building Kumer London

Cable offices Elwell London
Western Union Code

23

# WIRELESS:

## Popular and Concise

BY

## Lt-Col. C. G. Chetwode Crawley

R.M.A.

Member of the Institution of Electrical Engineers

Deputy Inspector of Wireless Telegraphy General Post Office

Secretary to the Wireless Telegraphy Commission for planning
the Imperial Chain Stations

# CONTENTS

Portions of this book which appeared in articles by the author in "Discovery," "Business Organisation and Management," and "Electricity," are made use of here by the kind permission of the Editors.

*Semper aliquid novi ex Aethere.*

(Always something new from the Ether)

# ILLUSTRATIONS

# Wireless: Popular & Concise

## CHAPTER I

### THE HISTORY OF WIRELESS

The history of wireless signalling, like most other histories, is as full of interest for the reader as it is of difficulties for the writer. The reader's first question of "who invented wireless" must necessarily bring forth a long, hedging explanation. A good commencement, however, may be made by coupling the foundations of wireless with the names of James Clerk Maxwell, of England, Heinrich Rudolf Hertz, of Germany, and Guglielmo Marconi, of Italy. Maxwell, in 1864, read a paper before the Royal Society, in which he proved the existence of the waves now used in wireless signalling; but it remained for Hertz to demonstrate, in 1887, their propagation, and it is in his honour that these wireless waves are often called Hertzian waves.

In passing, however, we should pause for a moment to review the work of David Hughes in this country (United Kingdom). Hughes, who made his name and fortune by inventing a printing telegraph apparatus in 1854, showed some experiments to several of our leading physicists in 1879 and 1880. In these experiments he demonstrated that an electric spark could affect a piece of apparatus, afterwards rediscovered independently and called a Coherer, at a distance of a few hundred yards. Hughes was really making use of wireless waves, but the experts gave him little encouragement, and his results remained unpublished and unappreciated for twenty years.

The results of Hertz's classical experiments were quickly taken up by scientific workers all over the world; but it was not till 1895 that Marconi, with intuitive genius, foresaw something of the great practical possibilities of wireless communication, and rightly, indeed, will his name be linked for all time with the invention of wireless telegraphy.

In 1896, Senatore Marconi came to England, where he was introduced to Sir William Preece, Engineer-in-Chief of the Post Office, at whose request he gave his first demonstration in this country. This demonstration was over a distance of only about a hundred yards, but by the end of the year the range had been increased to a few miles. It was on the 2nd of June of that year that Marconi lodged an application for the first Wireless British Patent. This patent was an application of Hertz's arrangements for sending and receiving, but with the important addition that in the case of signalling through intervening obstacles one terminal of the transmitter was to be joined to a vertical wire and the other to earth. This was the original Marconi Plain Aerial mentioned in a later chapter, and may be considered as the first great step in practical wireless transmission.

A host of other workers were by this time investigating the phenomena, which had been brought to the notice of the scientific world by Hertz, and was already being introduced by Marconi to commercial circles. Foremost amongst the scientists was Sir Oliver Lodge, who in 1897, the year after Marconi's first patent, himself patented arrangements of tuning which were a great advance on any suggestions in Marconi's patent, so far as the possibilities of signalling through interference were concerned. In these patents —Marconi's of 1896 and Lodge's of 1897—were, in fact, contained the essential principles of wireless signalling; Marconi's had the high antenna and the earth connection, the essentials for radiation; Lodge's had the electrical tuning of circuits, the essential for communication through interference. Lodge also suggested what is called a counter-capacity, a sort of lower aerial system, instead of the direct earth connection, and this is still largely used in one form or another.

In this country the possibilities of wireless for naval communications were grasped from the first by Commander Jackson, R.N. (now Admiral of the Fleet Sir Henry Jackson), who carried out a brilliant series of experiments in these early days, and by whose efforts the British Navy was the first of all navies to be equipped with wireless apparatus.

But it was not by any means in England only that advances were being made. France, as usual in scientific progress, was well up in the van. In 1892 Edouard Branly, who had rediscovered the Coherer action that Hughes had made use of twelve years before, produced the filings Coherer, which in the hands of Marconi was developed a few years later into the first practical device for receiving wireless signals. In the United States of America, Nikola Tesla and others were working on the lines of Marconi, but it was not till later that the full force of American inventive genius made itself felt.

In 1897, Marconi gave demonstrations of signalling over distances of ten miles or so to Government officials in England and Italy, and by the end of the year the first commercial station was being erected at the Needles in the Isle of Wight, the first paid message being sent by Lord Kelvin on the 3rd June, 1898. The use of wireless for ship communications in cases of distress was shown more than once during 1899, and from that year it was clear that the compulsory fitting of wireless in ships was simply a matter of time, though few then thought that the time would be so long as twenty years.

As the number of stations began to multiply so did the difficulties of interference become more and more apparent. Lodge's patent of 1897, already mentioned, was the first to show possibilities of improvement, and it was not long before other workers were hot on the trail. Tesla, de Forest, Fessenden in America, Braun and Slaby in Germany, Branly in France, Marconi in Italy and England, and Lodge in England, were the leaders in initiating that flow of wireless patents which has continued unceasingly ever since. Of all these early patents, Marconi's No. 7777 of 1900, known the world over as the "four sevens" patent, is the most famous, and has since been shown in the law courts, on more than one occasion, to be a master patent. It provides for the tuning to a common wave-length of the four main wireless circuits, the spark circuit, the sending antenna, the receiving antenna, and the detector circuit.

In October, 1900, Marconi commenced the erection of a trans-Atlantic station at Poldhu in Cornwall, and in December, 1901, he received signals from this station in St. John's,

Newfoundland. This achievement is recognized, and, indeed, was recognized at the time, as an epoch-making event, the inauguration of long distance wireless communications. There were many authorities who thought that such long distance working was impossible, but Marconi's genius and enthusiasm for practical results brushed aside all such theoretical deductions. He has always had, indeed, a disarming knack of first doing the job, and then letting others show why it is possible.

During all these early days of wireless, the messages were recorded as Morse signs, i.e., dots and dashes on a paper tape, but in 1902 Marconi invented an instrument called the Magnetic Detector, which was much more reliable and sensitive than the various types of coherer then in use. With this instrument the Morse signs were heard as buzzing sounds in a telephone receiver, instead of being printed on a tape. Other forms of magnetic detector had been previously invented, but Marconi's type was the most suitable; at any rate, it was the one adopted by the Marconi Company, and was successfully used for many years in this country. About the same time, in Germany and America especially, Electrolytic Detectors with telephone reception displaced the coherers, and these two, magnetic and electrolytic detectors, held the field until the advent of Crystal Detectors, good types of which were first introduced by General Dunwoody and G. W. Pickard in America in 1906. Crystal Detectors were more sensitive than the magnetic and electrolytic types, but were not so reliable, especially as the magnetic, and it was a long struggle before the crystals swept them finally from the field.

But even before the Crystal Detector was invented its ultimate supremacy was doomed by Dr. J. A. Fleming, of London, in his patent for a Valve Detector in 1904. The writer, amongst others, at that time witnessed a demonstration by Senatore Marconi, in which the detector was contained in a locked box. Remarkable results were shown, but the box remained locked throughout, much to the disappointment of the onlookers, and the mystery of what was known in wireless circles in the Navy as the "locked box of Poldhu" remained unsolved until Fleming's Valve patent was published some months later.

Fleming's Valve was not a better detector than the best Crystal type, but it was the foundation of the Triode, or Three Electrode Valve, which, first suggested by Dr. Lee de Forest in America in 1906, revolutionized wireless signalling not only for receiving purposes but also for transmitting.

This brings us back to developments on the transmitting side. Apart from those arising from the tuning patents of Marconi and Lodge, there were few important developments for about the next five years, when the spark gap arrangement was altered. The original type of spark gap, which consisted of two fixed metal balls, gave place to rotary gaps, which were much more efficient, especially for powerful installations, and in 1906 Max Wien, in Germany, suggested an entirely new design of gap which was developed into what is called the quenched gap. Either one or other of these types, rotary or quenched, is still used in all spark installations.

In this same year, 1906, V. Poulsen, of Denmark, demonstrated in London a new system of generating wireless waves, based on a discovery made six years before by W. Duddell in this country. Poulsen's patent was taken out in 1903, and consisted essentially of the substitution of an electric Arc in place of the Spark, by which means he demonstrated that the waves radiated were persistent, and, amongst other advantages, allowed of much sharper tuning than in the case of spark systems. This was the first so called Continuous Wave System. It was adopted slowly, almost grudgingly, in Europe, but was more quickly taken into favour in the United States, and today is in use at many large stations all over the world.

In 1911, Dr. R. Goldschmidt, in Germany, utilized a high-speed generator for the production of wireless waves, which were also of the Continuous Wave type. Many inventors had been grappling for years with similar arrangements, but Goldschmidt's invention was first in the field for high-power installations. Since then, several workers, notably, Alexanderson in the United States, and Bethenod and Latour in France, have designed alternator systems which are now in common use for high-power installations.

The only other Continuous Wave arrangement is the Valve system. We have already seen that de Forest in 1906 suggested the use of a Three Electrode Valve, or, as he called it, an Audion, for the reception of signals, the Two Electrode Valve having been patented by Fleming two years before.

For the next five years development seemed slow, but many eminent physicists and engineers were hard at work, and by 1914 it was apparent that the Valve had opened up a new era for wireless communication. The outbreak of war naturally gave a great impetus to the practical application of Valves. A host of names might be mentioned in connection with this war work, but if it were to be confined to one, there are few who would not give that of General Gustave Ferrie, of France.

Valves are now used universally for receiving purposes to the exclusion of all the older detecting devices, and by their use signals can be heard loudly or even recorded where, before, nothing at all would have been received. They are used extensively, also, as Continuous Wave transmitters, and are to be used for this purpose in the high-power stations of the Imperial Chain.

This short sketch of the history of wireless would be incomplete without some mention of the regulations which govern its use. The first International Convention for regulating wireless, or, as it is officially called, "radio," signalling dates from 1906. This Convention laid down the standard wavelengths to be used for commercial ship and shore communications, and the methods to be employed in calling up, answering, sending messages, accounting, etc., as well as the important rule that stations had to intercommunicate without reference to the types of apparatus with which they were fitted. The next Convention, signed in 1912, amplified the provisions of the 1906 Convention, and an international conference for drawing up a new Convention is expected to take place next year in 1923.

In addition to these international regulations, countries have framed their own rules for local control, as, obviously, chaos would result from the indiscriminate use of wireless signalling.

# CHAPTER II

## WIRELESS WAVES

IN wireless signalling, energy is transferred from the sending station to the receiving station without any transference of matter, in a similar manner to that in which light, heat, or sound is transmitted from one place to another. In each case the energy is transferred by wave motions in the intervening medium.

Now, light and heat are transmitted from the sun to the earth not only through the atmosphere, but also through millions of miles of empty or vacuous space, so that the medium for the propagation of light and heat, whatever it is, is not a solid, a liquid, nor a gas, but must be something which is present in a vacuum.

This infinitely tenuous, all-pervading medium is called the ether, and it has had to be endowed with so many intricate, and sometimes apparently contradictory, properties to fit in with all the known laws of the propagation of light and heat, that we need not consider them here. However, the assumption of the existence of the ether with these definite properties is a fair one to make until researches, such as those on the theory of relativity, give us, as perhaps they will, a clearer understanding in these matters.

In conjunction with such common phenomena as light and heat, we must naturally consider sound and we find that the energy which produces sound is also transmitted by wave motions of a medium, which in this case, however, is gaseous (as air), or solid, or liquid.

The difference in the media in the cases of sound and light can be readily demonstrated by connecting up an electric bell and its battery under the receiver of an air pump, so that the space surrounding the bell, enclosed in the glass receiver, can be exhausted as far as possible of air. It will be found that, as the air becomes more and more rarefied, the sound of the bell decreases, eventually ceasing altogether when the receiver is exhausted of air. The hammer, however, can still be seen striking the bell, which shows that light is traversing the exhausted space, whereas sound is not.

Sound is the sensation produced in the ear by oscillations to and fro of the medium, in this case the air, these oscillations having been set up by a vibrating body. A rough analogy to the transference of energy by sound can be made by considering a row of billiard balls in contact with one another, as little particles of the medium. If the ball at one end is struck by another ball, the ball at the other end moves off, showing that energy has been transferred from one place to another by oscillations of the medium backwards and forwards in the line of propagation, without any transference of matter having taken place. That is to say the billiard balls which are between the two end ones are supposed to oscillate to and fro, coming to rest in the same positions as before, and the fact that they will not conform exactly to these movements in practice need not be allowed to spoil our analogy.

The speed at which energy is thus transferred in the case of sound waves depends primarily on the medium; in air, for instance, the velocity is about eleven hundred feet a second, and in water it is about four times as great.

Sound waves are produced, therefore, by oscillations in the medium to and fro in the line of propagation, but waves may be produced in a medium by oscillations taking place in directions across the line of propagation. Take, for instance, the case of the waves produced by dropping a stone into a pool of water. The energy imparted to the water by the stone is radiated in the form of circles of waves, the circles increasing in size, and the heights of the waves becoming smaller, until they reach the edges of the pool, or die away altogether. Now, if some corks are floating in the pool, the energy imparted to the water by the stone, and propagated by the waves, will cause the corks to bob up and down, thus showing that the little particles of water, which are in this case the medium, merely oscillate up and

Generators, Northolt Station

[*To face page* 22.]

down, and by doing so, convey energy in the form of waves, the corks nearer the original disturbance receiving more energy than those further away.

The stone, by disturbing the water at one place, has thus transferred energy to other places by waves in the medium, the water, without any transference of matter taking place.

Similarly, if one end of a piece of rope is fixed, and the other end is rapidly moved to and fro by hand, energy is transferred by wave motions in the rope from the hand to the place where the rope is attached without any matter being transferred from the one end of the rope to the other.

In both these cases the wave motion by which energy is transferred is produced by oscillations of the little particles which make up the medium—that is, the water or the rope, to and fro at right angles to the line of propagation of the energy. A wave length is the distance separating the crests of adjacent waves, and it will be apparent, by observing the motions of the rope, that the more rapid the oscillations the shorter are the wave lengths, and vice versa.

It is clear, therefore, that sound waves are quite different from light and heat waves; sound being produced by oscillations to and fro, in the direction of propagation of a solid, liquid, or gaseous medium: light and heat by oscillations, across the line of propagation, of an infinitely tenuous medium, called the ether. In the case of sound, the pitch of the note heard depends on the rate of oscillation, or the wave-length; in the case of light, the colour seen depends on the rate of oscillation, or the wave-length.

Now if energy is transferred at a speed of eleven hundred feet a second by oscillations of such a dense medium as air, it is to be expected that energy will be transferred at an enormously greater speed by oscillations of such a medium as the ether, and this is so, as in the latter case the speed is about 186,000 miles a second—that is, nearly eight times round the earth in a second. For all practical purposes, such a speed for distances over the earth's surface is instantaneous, though it is, of course, very far, indeed, from instantaneous when considering the stupendous distances dealt with in astronomy.

If the oscillations of the ether are at certain rates, they affect the eye and produce the sensation of light; if at certain other rates, they are received not as light, but as heat.

If the ether is set in oscillation at very much slower rates than can be detected as light or heat, the ether waves produced are those which are used for communication by wireless telegraphy. These wireless waves are often called Hertzian Waves, after Hertz, the German physicist, who in 1887 published the results of his experiments on the production and properties of these waves, results which confirmed theoretical deductions published in this country by Clerk Maxwell in 1864.

The transference of energy by all these ether waves, light, heat and wireless, takes place at the same rate, viz.: 186,000 miles a second; but the rates of oscillation of the ether, on which depend the lengths of the waves, differ very much in each case.

The highest known rates of oscillation in the ether produce Röntgen, or X rays; then come the actinic waves, which are used in photography, and then the visible light rays, at rates of oscillation from about 800 billions a second for violet light, down through indigo, blue, green, yellow and orange, to about 400 billions a second for red. A combination of all these colour rays produces white light—i.e., sunlight. Below these rates of oscillation are the heat rays, and then the rates of oscillation, about 6 millions to 10,000 a second, which produce the Hertzian waves now used for wireless signalling.

These wireless waves are therefore very similar to heat or light waves, and travel at the same speed, the main difference being that the rates of oscillation of the ether which produce them, though so exceedingly rapid, are far too slow to produce the sensation of heat or light; in other words, these wireless waves can be neither felt nor seen, and they cannot, of course, be heard, as they have no connection with sound waves.

Signalling by an unscreened flashing lamp, such as a warship's mast-head lamp, may be taken as an example of signalling by light waves, flashes of short and long duration representing the dots and dashes of the Morse Code. These flashes consist of trains of ether waves which radiate in all directions at a speed of 186,000 miles a second, the oscillations getting weaker and weaker as the distance from the lamp increases, until at last they have dissipated their energy and die away altogether.

The eye can be used to receive these signals, and the message in the Morse Code can be read simultaneously from all directions by any number of people within range, provided that there are no screening effects caused by opaque bodies intervening between their eyes and the lamp.

Wireless signals are sent in a very similar manner by short and long trains, or series of trains, of ether waves, which travel outwards from the sending station in all directions at this same speed of 186,000 miles a second. These trains of waves actuate, as dots and dashes according to their duration, the wireless receiving instruments at any number of stations which are within range and are adjusted to respond to the particular wave-length used.

These long ether waves are able to pass readily through bodies which are opaque to light or heat waves, unless such bodies consist of, or contain, conductors of electricity, such as metals, in which case the waves produce currents of electricity in the conductors, and so part with some of their energy.

Screening effects in the case of wireless waves are therefore very different from those which obtain in the case of light waves; for example, buildings, etc., opaque to light waves, may present no obstacle whatever to the passage of wireless waves; even mountains, and, indeed, the obstacle caused by the earth itself, do not affect these waves in the same way as they affect light waves.

For instance, there is no difficulty in signalling by wireless waves across the Atlantic, whereas it is impossible to signal over distances of more than a comparatively few miles by light waves, owing to the earth imposing, due to the curvature of its surface, an impenetrable obstacle to such waves. In the case of the Atlantic Ocean this obstacle would represent a mountain some hundreds of miles in height, and the waves do not actually pass through this obstacle, but may be considered as gliding over the surface. Like light waves, they are subject to absorption, to reflection, and to other forms of bending, as refraction, on account of which mountains or other high land close to a wireless station may produce serious screening in that direction, whereas the same obstacle at a distance may have very little or even no appreciable effect. Short waves are much more affected by this land screening than are long waves; but, in any case, it is advisable to avoid choosing a site for a wireless station close to higher ground in a direction towards which communication is required.

As a matter of fact, very little is known about screening, other than what has been found in practice. There is no doubt that a wireless station placed close under a mountain will not be able to communicate as far in directions over the mountain as in other more open directions; but what reduction of signalling range will be produced by the mountain can only be guessed. Mountains at a distance have very much less effect; for instance, the Alps may have little effect on signalling between England and Egypt; but that they have some effect is equally probable. Then there are the effects produced by river valleys, which seem to assist the passage of the waves. Ships in the Mediterranean, for example, have often noted an increase in the strength of signals from English stations when the line between the station and the ship coincided with the general direction of some great river valley in Europe.

In this connection it must be noted that the path of the waves from a sending to a receiving station over the surface of the earth is the shortest possible; that is to say, the waves travel along the great circle between the two stations, a great circle being a line drawn round the earth, dividing it into two equal portions. Again, there are rather obscure effects produced by the path of the waves being diverted by long stretches of coast-lines; for instance, a ship receiving signals from a coast station will sometimes find that the signals weaken as the ship approaches the coast at a point separated from the station by a stretch of coast-line, though it may actually be closer to the station than before.

So far we have considered the wireless waves as being radiated in all directions from the sending station; but it is necessary to explain shortly how it is that they travel over the surface of the earth. This is due to the fact that the space in which the waves travel is a spherical shell, the under side being the earth's surface and the upper side a layer of air probably about thirty miles up from the earth. The lower parts of the waves travel along the earth's surface, as it is an electrical conductor; the upper parts along the layer of air. This layer is a much better conductor than the earth, so that the upper parts travel more rapidly than the lower. The waves are thus bent with the upper parts deflected towards the earth, the effect being very similar to the manner in which the sun's rays after sunset, due to bending effects in the atmosphere, shed light on the earth.

The effects produced by different electrical states of the atmosphere are not yet fully understood, but some well-established facts are certainly due to the bending and absorption of the waves. The energy of the waves is absorbed more readily during the day than during the night, or, in other words, signals are stronger at night than in the daytime, and this is especially noticeable in the case of short waves. Generally speaking, short waves only can be installed in small sets, such as those in ships, aircraft, or small land stations, and it is at such stations that "freak" distances are so often experienced. For instance, it is no uncommon event for a ship fitted with an installation which can normally send signals to another ship up to, say, two hundred miles by day, being able to send signals to the same ship at distances of, say, twelve hundred miles by night. Ranges normally obtained in one part of the world differ, too, from those obtained in another, quite apart from the effects of interference from electrical disturbances in the atmosphere; for instance, ships can communicate over greater distances in the Pacific than in the Atlantic Ocean. With the longer waves, which are only suitable for transmission from large stations, the diminution of range in the daytime is very much less; in fact, with the longest waves now used there is often no diminution. These long waves are also better for signalling over land, mountains, etc., than the short ones. There is, of course, no hard and fast line of demarcation between long and short waves, but wave-lengths of over, say, five thousand metres may be classed as long waves. The day and night effect is very marked when a station in daylight is communicating with a station in darkness. In such a case conditions are very bad for intercommunication, and are probably at their worst when it is sunrise at the one station and darkness at the other.

The whole matter still needs much investigation, but it is, at any rate, certain that at sunrise and sunset signals are not at their normal strength, and usually they are weaker.

For signalling by wireless, the sending station does not transmit a number of ether waves of comparable strengths and of various lengths at the same time, as is the case when signalling by light, but makes use of one wave-length only for that particular transmission.

All wireless communication is not, however, carried out on the same wave-length, as, if it were, the receiving stations would receive a jumble up of all the messages being sent on that wave-length from stations within range, instead of the particular message which it was desired to receive. When the reception of a message is thus interfered with by other messages being sent at the same time, the message is said to be "jammed." This jamming of a message may also be caused by stray ether disturbances in the atmosphere itself. These disturbances are called "atmospherics" or "strays" and are specially strong when thunderstorms are in the vicinity, lightning itself causing very violent atmospheric disturbances. Atmospherics often cause great interference with wireless communications, and in tropical countries are sometimes so strong as to prevent any communication at all for hours at a time. They are strongest, as a rule, at night and in the afternoon. In more temperate latitudes, as in this country, atmospherics are usually at their worst during the night in summer-time, but are seldom so troublesome as those experienced in warmer climates.

The worst of these disturbances is that they are not confined to any particular wave-length, but interfere to a greater or lesser extent with any wireless signalling that is taking place within their range of action. Generally speaking, however, signalling on long waves is more troubled by atmospherics than signalling on short waves, and this is fortunate, as stations which use long waves are large powerful stations, which can more easily arrange for sending out wireless signals strong enough for reading at the receiving stations, even through the false signals caused by atmospherics. This jamming by atmospherics is the bugbear of wireless, and a very serious bugbear it is.

It has been recognized from the beginning that the really great problem of wireless is how to minimize the effect of atmospherics at the receiving station. The periodic announcements of the invention of infallible atmospheric stopping devices may be ranked with those of the sea-serpent; but still, there is no doubt that every year does see substantial advances made in the long-drawn-out struggle with this great hindrance to world-wide wireless communication.

The choice of the wave-length to be transmitted depends on various considerations.

At large, powerful stations, which are designed to send signals over great distances, long waves are transmitted, because, given sufficient power, long waves travel better than short waves, and high-power transmitting apparatus can produce long waves more efficiently than short waves. At small stations the use of high power is impracticable, and the best results can be obtained by using short waves. For instance, a wave-length of 600 metres is that normally used for transmission from ships, and is an efficient wave for use in installations of a size suitable for fitting in ships. Wave-lengths as long as 24,000 metres are being used at large stations, and as short as 100 metres at small stations.

There are, however, other important considerations to be taken into account in fixing the exact wave-length for any given communication, and the most important of all is that of jamming. If a station is receiving a message on, say, a wave-length of 600 metres from another station, the message will be jammed, as mentioned above, by any other station within range, which is sending a message on the same wavelength. It is essential, therefore, to use different wave-lengths for different communications in the same zones, and it is this aspect of wireless, which necessitates international control to an extent far greater than is the case with any other means of communication. We can only hope that this necessity for international intercourse, involving, as it must, a give-and-take policy for the common good, has advantages, which, to some extent at any rate, counterbalance the disadvantage of this great technical difficulty in the organisation of wireless communications.

# CHAPTER III

## HOW WIRELESS MESSAGES ARE SENT

WE have seen that wireless signals are sent by radiating ether waves from the transmitting station for short and long periods corresponding to the dots and dashes of the Morse code, in which every letter or figure is made up of these symbols, just as light signals can be sent by the short and long flashes of an electric lamp. In each case the waves consist of oscillations of the ether, and travel at the same speed of 186,000 miles a second, but the rate of oscillation for wireless waves, not higher than a few million a second, are far too slow to produce the sensation of light which requires oscillations at the rate of some hundreds of billions a second.

The first question is, therefore, how can the ether be set in oscillation at rates of between a few million and a few thousand a second, the rates required for the production of these wireless or Hertzian waves, and the answer is that this is done by currents of electricity oscillating to and fro at these very high speeds. Such currents are called oscillatory currents, and can be produced by various means.

The first method used for producing oscillatory currents was to allow a condenser which had been charged up with electricity to discharge across an air gap (see Fig. 1), when it was found that the flow of electricity, that is, the current, on discharge oscillated to and fro at a rate suitable for producing wireless waves. A condenser is a piece of apparatus made up of two electrical conductors, such as metal plates, separated from one another by some electrical non-conductor such as glass or air, and can be charged up with electricity by connecting a source of electricity across the plates.

30

Suppose, then, that the plates are connected together by a piece of wire with a gap in it, as at the spark balls in Fig. 1,

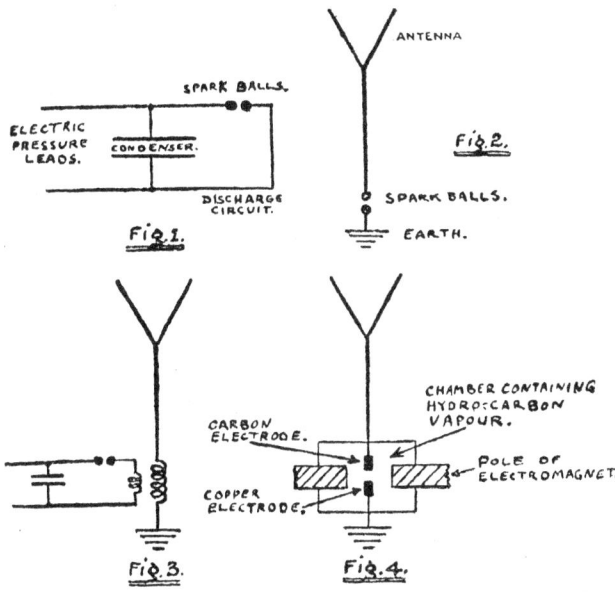

a flow of electricity will take place from one plate of the condenser across the gap, where it forms a spark, to the other plate. This flow of electricity, or current, will oscillate to and fro between the plates, across the gap, until its energy has been expended, just as a pendulum if pulled aside, and released, will oscillate to and fro until its energy has been expended, and it comes to rest.

This was the method by which Joseph Henry in 1840, in the United States, produced currents which he showed were oscillating at these very high speeds. It was the method used by Hertz in his famous experiments at Karlsruhe in 1886 when he demonstrated that these currents produced the ether waves now used for wireless signalling, and, lastly, it was the method first used by Marconi, who, in 1895, in Italy, produced the oscillatory currents in an elevated wire, and thus put into a practical form for signalling an arrangement which was until then only a laboratory experiment.

The elevated wire, called the antenna or aerial, used by Marconi has taken many different forms, but all transmitting antennae consist essentially of a wire, or a system of wires, with the lower end connected to the sending apparatus, and the upper end suspended from high masts. Usually the upper portion, or roof, of the antenna consists of a network of wires suspended from two or more masts, this network being connected to the transmitter below by a wire or number of wires, called the feeder (see Fig. 5). The roof is suspended from the masts by electrical insulators, made of some non-conducting material such as porcelain, as it is essential that no part of the antenna be connected electrically to a mast or any other body in contact with the earth; in other words, the antenna is electrically isolated in space so that the oscillatory currents which flow in it cannot leak away to the earth.

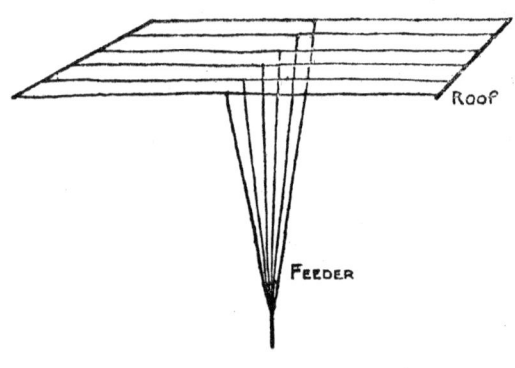

Roof

Feeder

Fig. 5.

The height at which the roof is suspended depends, of course, upon   circumstances, but, generally speaking, the higher the better.

In ships, a height of over about 120 feet is seldom possible, but at land stations it is the cost of high masts or towers that is usually the governing factor, as increase in cost is out of all proportion to increase in height. At some of the most powerful stations, masts or towers of between 800 and 1,000 feet high are in use, and heights of 400 or 500 feet are quite common for stations of medium power.

In the case of aircraft, the antenna usually consists of a single trailing wire let down when in the air, and kept wound up on a drum when not in use.

In Marconi's first transmitter (see Fig. 2), the spark gap consisted of two metal balls separated by an air space of about an inch, one ball being connected to the antenna and the other to "earth." The two conductors and the non-conductor which formed the condenser were made up, therefore, of the antenna, the earth and the air respectively. This method, often referred to as "Plain Aerial," produces wave trains which are very highly damped, that is, which die away very rapidly.

A receiving station, though specially arranged, or tuned, to respond to waves of one particular length, will respond nearly equally well to waves of other lengths, if the wave trains are very highly damped. This means that a message sent by Plain Aerial on a particular wave length will be received at the same time as messages which are being sent on other wave lengths, so that an unintelligible jumble of dots and dashes will be heard at the receiving station. The general use of Plain Aerial transmitters would thus produce much jamming of messages, and for this reason its employment for ship communications, except in cases of emergency, is prohibited.

The Spark System is, however, used at the present time for practically all ship communications, but the oscillatory currents are induced in the antenna from a separate circuit (see Fig. 3), instead of being produced by directly charging and discharging the antenna as in the Plain Aerial arrangement. Spark transmitters of this nature are said to be tuned, as the rate at which electric oscillations take place in the condenser circuit is adjusted, or tuned, to be the same as that in the antenna circuit, so that the oscillatory discharge current from the condenser may induce a large oscillatory current in the antenna.

In Plain Aerial, the spark across the gap electrically connects the antenna to "earth" when the oscillations are taking place, and in a tuned system, the bottom of the antenna is permanently joined to "earth," for it must be remembered that the earth plays a very important part in the propagation of wireless waves which pass over its surface. In ships, the "earth" consists of the hull of the vessel, which is in electrical connection with the sea, except in the case of wooden ships, where more elaborate arrangements for providing a good electrical "earth" must be made. At shore stations, the "earth" usually consists of metal plates and wires buried in the ground, or a network of wires in contact with the ground over a large surface. In aircraft, metal portions of the fuselage are used as an "earth."

The original form of gap mentioned above, consisting of two fixed metal balls, is no longer used, but has been superseded by various arrangements designed principally with the object, not only of giving stronger radiation, but also of producing in the receiving instruments signals which have a distinctive musical note. The advantage of signals having a distinctive note when heard at the receiving station is that it is easier to read the messages through the gruff or crashing noises produced in the receiver by electrical disturbances in the atmosphere, or through other messages which, though heard, produce notes of different pitch. It is this advantage of being able to differentiate between different notes, and so read a message through interference, that makes reception by sound much more efficient when there is interference, than reception by instruments which record the message on a tape. Such instruments will record a mere jumble of dots and dashes whenever interference is received, irrespective of whether the interference would make the same sound in a telephone receiver as that which would be made by the signal required.

There are three main designs of spark gaps which give this advantage of producing signals that can be read through atmospheric disturbances provided these latter are not very strong. These gaps are of the fixed type with an air blast playing on the gap, or of the revolving type, or of the quenched type. All are designed with the object, amongst others, of preventing an electric arc, similar to that in an arc lamp, being formed across the gaps, as this arcing prevents the production of the musical note, and is undesirable in other ways.

In the air-blast type, the blast of air is used to blow out the arc. In the revolving type, the gap is quickly opened out by having a metal disc fitted with protruding metal studs, revolving between two fixed studs, so that any arc which tends to form is extinguished by being rapidly elongated between the fixed studs and the moving studs. In the quenched type, the spark gap consists of a number of metal discs connected in series with one another, their surfaces being separated from each other by only a few millimetres. This arrangement quenches the spark in the condenser circuit after a very few oscillations, and this tends to prevent arcing at the gap.

The revolving and the quenched types are the only ones now in common use, the former being first introduced by the British Marconi Company, and the latter by the German Telefunken Company. These are the two types of gaps used in ship installations, and both ensure that the notes of signals when received by sound are of much higher pitches than those produced by atmospheric disturbances.

The ranges over which signals can be transmitted from ships depend primarily on the energy of the ether oscillations produced by the oscillatory currents in the aerial, the more powerful the currents the greater being the range. The sort of range by day over which ships may be expected to communicate with each other, using spark systems, is about 200 miles, but this distance will be greater in the case of large ships where more efficient antennae and greater electric power are available, and less in the smaller ships where the reverse conditions prevail. By night, much greater ranges are usually obtained, as the energy radiated is not dissipated so rapidly as in daylight.

In addition to having a spark transmitter, many of the larger passenger ships are now fitted for long-range work, about 1,000 miles by day, with the Valve System of transmission.

A Three Electrode Thermionic Tube, or, as it is frequently called, Valve, consists essentially of a vacuum tube, similar to an electric light bulb, but having two separate metal elements inside the bulb with electrical connections to the outside, in addition to the filament with its outside connections. (See Fig. 8.)

The plate may consist of a cylindrical metal plate surrounding the filament, and the grid of a cylindrical wire mesh between the filament and the plate. If the filament is made incandescent, as in the case of an ordinary electric light, and an electric battery is suitably joined between the plate and filament, a current will flow from the plate across the space inside the bulb to the filament, passing through the grid. The strength of this current will depend on the battery power used, but can also be varied by applying electrical pressure to the grid. This pressure acts, therefore, like an electrical tap controlling the flow of current through the apparatus.

If oscillatory currents are started in the antenna, and are arranged to vary the electrical pressure on the grid, they may be made to cause larger currents in the plate circuit, from which energy can be transferred to the antenna, and the oscillatory currents there made stronger. This series of actions is then repeated till the antenna is set into strong oscillation, the apparatus thus acting as a generator of oscillatory currents.

The Valves now used for transmission are made of glass or silica, the glass, ones being about the size of a child's football and the silica ones rather longer and thinner. About sixty or twenty would be required for a really large installation, according to whether they are of glass or silica respectively. These thermionic valves have only been developed for transmitting purposes since 1916. At first they were used principally in aircraft, but are being fitted now in large ships, and are also being rapidly developed to handle the high power required for long-distance shore stations.

In this country, the Post Office station at Devizes is fitted with Valve transmission for long-range communication with the specially fitted ships; the Post Office Coast stations, of which there are twelve, being fitted with Spark transmission for normal ship working.

An arc transmitter as fitted at the Oxford and Cairo Stations.

[*To face page* 36.

The Spark System is now used in very few high-power stations, most modern stations being equipped with either Arcs or Alternators, though a few large stations have recently been fitted with Valve transmission, the system recommended for use in the Imperial Chain Stations.

The electric arc was first introduced as a wireless transmitter by Poulsen, at Copenhagen in 1903, the arrangements used being based on the results of Duddell's experiments in this country, published in 1900. The arc is arranged to take place in a magnetic field in an atmosphere of hydrocarbon vapour, usually between copper and carbon electrodes (see Fig. 4). The antenna and the earth are, as a rule, connected up to the electrodes just as a Plain Aerial spark transmitter, but the waves radiated by this arrangement are persistent and have not, therefore, the disadvantages of those radiated by the Plain Aerial spark.

The Post Office has erected a high-power Arc Station near Oxford and a similar one near Cairo for inter-communication, and there are numerous large Arc Stations throughout the world. The Arc System is also quite suitable for use in ships, and is being fitted, like the Valve, in many large ships for long-range work.

The first really successful arrangement for employing Alternators for high-power wireless transmission was devised by Professor Goldschmidt, in Germany in 1911, the oscillatory current in the antenna being generated through suitable circuits by an alternating current machine run at a high speed. Very high-speed machines are obviously not suitable for ships or aircraft, but are now being used in many high-power stations.

The Valve, Arc and Alternator all radiate a continuous stream of waves so long as the signalling key is pressed, and are usually alluded to as Continuous Wave transmitters. With them, much sharper tuning can be obtained than with Spark transmitters, which means that different signals sent on different wave-lengths, though the wave-lengths selected may be close together, do not interfere with one another at the receiving end. Greater distances can also be obtained by Continuous Wave systems compared with Spark systems of the same power, and in fact the former are more efficient in every way.

In telegraphy, when the signalling key is pressed for, say, a dot on a Spark system, trains of waves are radiated, and at the receiving station each train is made to affect the telephone receiver, the pitch of the note heard depending on the number of complete trains a second. If the key be pressed for a longer period, a dash with the same pitch of note is heard, and if the key is not pressed, there is no radiation and nothing is heard. In a Continuous Wave system, the pitch of the note of the signals can be controlled at the receiving end, an obvious advantage when trying to read through interference from other signals.

In telephony, where continuous wave systems are essential, a continuous stream of waves is modulated by the voice, as explained in a later chapter.

Every commercial station is allotted a call signal, which consists of a group of three or four letters, and these call signals, along with other details of the stations, such as the wave-lengths to be used, are published through the medium of an International Bureau at Berne. The calling station transmits the call signal of the station with which it wants to communicate and gives its own call signal. The other station, if it receives the call, replies similarly, and communication is established. Other stations within range standing by to receive on that same wave-length can read the messages if they wish, but they cannot themselves communicate on that wave-length, as they would be jammed as well as cause jamming of the original communication.

At most small stations, including all ship and aircraft installations, the manipulation of the signalling key is carried out by hand at a speed of about twenty words a minute, about thirty words a minute being the maximum speed attainable with hand working, but at the larger stations automatic transmission at speeds of between fifty and two hundred words a minute is the normal arrangement. In automatic transmission the messages are first punched up on a paper tape, two small holes one above the other representing a dot if the lower one is directly below the upper, and a dash if it is a space further on to the right. This punched slip is then passed through an apparatus by clockwork at the speed desired, small metal rods coming up through the punched holes and actuating electrical contacts through which the signalling key is worked electrically.

It is difficult to give a useful comparison of the Valve, Arc, and Alternator systems, as so much depends on circumstances. For high-power work the Valve has not yet been used to nearly so great an extent as either of the others, but the fact that experts in this country recommended in 1920 that it should be fitted in the large Imperial Chain Stations augurs well for its future success. For small installations it has been well tried out, and has proved satisfactory, but its performance and cost of maintenance in large sets are still somewhat unknown quantities. The Arc was first in the field for high-power sets, and is generally admitted to be the simplest of the three. The others, however, radiate what may be called a purer wave, and for that reason are better adapted for telephonic transmission. The Arc and the Alternator have both been well tried out as high-power transmitters. The simplicity of the Arc, which means easy maintenance, and the fact that the wave-length can be easily changed with little alteration in efficiency, fits in with some conditions, but on the other hand the fact that the Alternator is really just a normal development in electrical machinery, and transmits a purer wave than the Arc in its present form, makes it more suitable for others. The Valve is usually claimed to be the most flexible arrangement, admitting of easy changes in power and wave-length if desired, and ready replacements by the substitution of new valves in case of breakdown.

# CHAPTER IV

## HOW WIRELESS MESSAGES ARE RECEIVED

THE ether oscillations which form the waves used in wireless telegraphy are produced, as we have seen, by currents of electricity oscillating backwards and forwards at very high speeds, of the order of thousands and millions a second. These currents, called oscillatory or high frequency alternating currents, are caused to flow in an elevated wire or system of wires called the antenna, from which the wireless waves are radiated into space in all directions. Energy is thus transmitted from the sending station at the same speed as the energy which produces the sensation of light, that is, at about 186,000 miles a second.

These wireless waves, which are produced by oscillatory currents, have the property of producing exactly similar currents in any electrical conductor which they meet, and it is this property which is made use of in the reception of wireless signals.

In wireless telegraphy, by means of a signalling key at the transmitting station, series of ether waves are sent out for shorter or longer periods of time, that is, as dots or dashes, and produce oscillatory currents which continue to flow for the same periods in any conductor which they meet. At the receiving station, as at the sending station, the conductor takes the form of a system of wires, and to read signals in the Morse code the currents produced in the receiving aerial are made to actuate electrical apparatus in such a way as to produce the dots and dashes as marks on paper, or as sounds in a telephone receiver.

In wireless telephony, the procedure is very much the same except that the oscillatory currents in the transmitting aerial, and consequently the waves radiated, are modulated in strength by the voice in a similar way to that in which the currents are modulated in land line telephony. The waves, so modulated, produce in the receiving aerial exactly similar currents, which actuate apparatus for producing the voice sounds in a telephone receiver.

The strength of the currents in the receiving aerial will, of course, depend on the distance away of the sending station, as the ether oscillations become weaker and weaker as the distance over which the waves travel increases. A powerful sending station is one in which large currents are produced in the aerial, these currents producing strong ether oscillations, which allow of the resulting waves being detected by receiving stations at great distances. Signals from the most powerful European stations can sometimes be heard even in New Zealand, but efficient communication at all times of the day and night over these very long distances is not yet a practical proposition.

The most powerful transmitting stations in this country are the Marconi Company's station at Carnarvon, and the Post Office station at Leafield, near Oxford, signals from both of which are frequently received in Australia. A still more powerful one will be erected shortly by the Government in connection with the Imperial Chain of stations, and will be able to work for considerable portions of the day with Australia.

In these large stations, separate antennae are used for sending and receiving; in fact, the receiving station is usually some miles away from the sending station, so as to allow of arrangements being made for duplex working, that is, for messages being transmitted to a distant station at the same time as messages are being received from that or another station. In some of the latest designs of station the most sensitive possible receiving instruments are used, and the receiving aerial is quite a small affair in the form of a loop or frame, as by using such arrangements it is easier to eliminate interference from the sending station, so facilitating duplex working, and at the same time avoiding the great expense of high masts. In the case of Carnarvon, the receiving station is at Towyn, some thirty miles away. The operating is also carried out at Towyn, the signalling key being connected electrically by wires to the transmitting apparatus at Carnarvon. Similarly, the receiving station for Leafield is at Banbury, where the signalling key for operating the apparatus at Leafield is also installed.

Indeed, this splitting up of a station into separate sending and receiving units, for the purpose of being able to work duplex, is now carried out at all point-to-point stations where the amount of traffic, i.e., the number of telegrams dealt with, justifies the extra expense. In ships and aircraft the amount of traffic is not sufficient to call for duplex working, and if it did, it would be very difficult to arrange in ships, to say nothing of aircraft, owing to the difficulty of obtaining a sufficient distance between the sending and receiving installations. The amount of traffic carried out by ships is not even sufficient to justify the installation of high-speed automatic apparatus, so the time has certainly not yet come when the possibilities of duplex working need be seriously considered. At present, the amount of traffic to and from aircraft is very small, but rapid working is important, owing to the great speed at which these craft travel away from, or towards, land stations, and for this reason, amongst others, wireless telephony is very suitable for aircraft communications. In ships, aircraft, and other small stations not fitted for duplex working, the same antenna is used for transmission and reception, being switched over by the operator to the transmitting or receiving apparatus as required.

When receiving signals from a distant station, the electric currents produced in a receiving aerial by the waves radiated from the distant station are extremely small, and apparatus must be used to enable these minute currents to release sufficient energy for working some sort of recording arrangement, whether it be one for recording sounds in a telephone receiver or marks on a paper tape. The apparatus first used was the coherer, which consisted of a sealed glass tube in which were two metal plugs, separated by a small space containing metal filings, as shown

**Coherer**

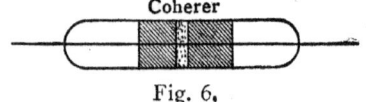

Fig. 6,

in Fig. 6. Oscillatory currents have the property of very greatly reducing the electrical resistance of the little heap of filings, which are then said to have cohered together, and in this state can allow a current from a cell to flow through them and work the recording apparatus.

This coherer action was enunciated by Prof. Branly in France in 1890, and was adopted for wireless signalling purposes in this country by Sir Oliver Lodge, who called the apparatus a "Coherer." In 1895 it was used by Professor Popoff in Russia as a detector for aero-electrical discharges, and in 1896 was in use by Signor Marconi as a detector in practical wireless working. This filings coherer, along with a few other detectors whose working also depended on coherer action, held the field for some years before being supplanted by more sensitive arrangements.

It was first succeeded by Marconi's magnetic detector, which was introduced in 1902, and quickly followed, though not supplanted, by various electrolytic devices, of which the first really useful detector was introduced in 1903, in the United States, by Professor Fessenden.

The magnetic and electrolytic detectors could only be used in conjunction with a telephone receiver, that is, the message was not printed on a tape as dots and dashes, which was the case with coherer devices, but was read by ear as buzzing sounds in a telephone receiver, the duration of the sounds signifying the dots and the dashes of the Morse code. At first this fact of not being able to have the message printed on a tape was thought by those not directly acquainted with practical working to be a great drawback, as it was said that there could be no certainty that the operator wrote down the message exactly as received, whereas with the dots and dashes printed on a tape there could be no doubt of the matter. The writer, indeed, well remembers many forcibly expressed opinions from that point of view when it was first suggested in the Navy that aural reception should supplant recorded signals. However, the magnetic and electrolytic receivers were much more sensitive and reliable than the filings coherer, and it was soon found that this advantage, as well as that of the operator being able to read messages to a certain extent through atmospheric and other interference, far out weighed the disadvantage of not being able to check what was received by reference to a printed record.

The telephone receiver consists of an electro-magnet fixed close to a circular metal disc supported round the

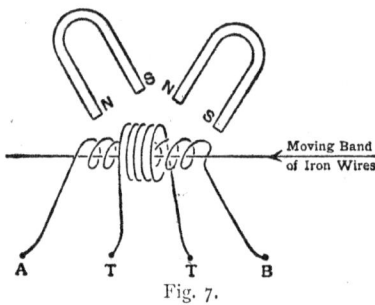

Fig. 7.

circumference. An alternating or pulsating current flowing through the coils of the magnet alters the magnetic field in such a way as to vibrate the disc and produce sounds in the ear. The instrument is a sensitive current detector, but cannot respond to currents which oscillate at such very high speeds as those produced by wireless waves, so that some device must be introduced which will utilize the energy of these currents in such a way as to produce currents suitable for operating the telephone.

In the Magnetic Detector, see Fig. 7, the minute oscillatory currents received are passed through the coil A B and alter the magnetic field so as to produce currents in the coil T T, suitable for working the telephone receiver joined up to the ends of the coil.

In the electrolytic type of detectors, as in coherers, the oscillatory currents reduce the resistance of the circuit in such a way as to allow the passage of a current suitable for working a telephone receiver. The detector consists of two dissimilar metals, such as platinum and lead, immersed in an acid solution, the platinum having only a small point, the lead a large surface, in contact with the liquid. When at rest, a non-conducting film forms on the point, causing the circuit through the liquid from one metal to the other to have a high resistance.

Transmitting apparatus, Stonehaven Station.

A valve receiving set.

[*To face page* 44.

Oscillatory currents disturb this film, and the resistance decreases, so that the current for working the telephones can flow through the circuit.

In 1904, Professor J. A. Fleming in this country invented a detector which he called the Oscillation Valve. This consists of an incandescent lamp with the addition of a metal plate, or cylinder, inside the bulb, connected to a wire sealed through the glass. When the lamp is glowing, the space between the filament and the plate is electrically conductive, but in one direction only, that is to say, it offers a very great resistance to a current in one direction but very little resistance to a current in the opposite direction. If, therefore, an oscillatory current is applied, it becomes for all practical purposes an unidirectional pulsating current which can be used for working a telephone receiver. The device acts in fact as a valve for the oscillations, allowing current to flow in one direction but not in the other.

Soon after the introduction of the Fleming Valve, another form of receiver, called the Crystal Detector, was introduced. The first practical form was used by General Dunwoody in the United States, and consisted of a crystal of carborundum in contact with a piece of metal. Various combinations of crystals and metals give a rectifying effect, that is, they convert an oscillatory current, applied across the contact, into a form suitable for operating a telephone receiver.

These Crystal Detectors are still occasionally used in ships and small stations, but have really been superseded by Thermionic Tubes or, as they are often called, Valves. The form commonly used is one with three electrodes, filament, grid and plate, just as in the case of transmitting valves, except that they are very much smaller, being about half the size of an ordinary electric light bulb.

The Three Electrode Thermionic Tube, small and simple as it appears to be, has revolutionized the whole outlook of wireless. The idea of using Vacuum Tubes for wireless reception dates from 1904, when, as mentioned above, Prof. Fleming in this country used a vacuum tube with two elements, a filament and a metal plate, inside the tube, but the introduction of the third element, the grid, was first suggested a couple of years later in the United States by Dr. de Forest, who called the apparatus an Audion.

No other device can compare with a Thermionic Tube for the reception of weak signals, and it is as a receiver that it is now in universal use. Its employment as a transmitter came later, and in that sphere it has serious competitors, the Alternator and the Arc, as mentioned in the last chapter.

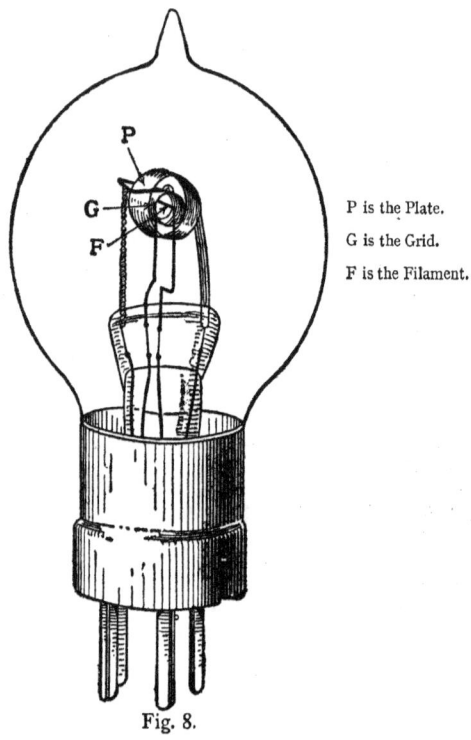

P is the Plate.

G is the Grid.

F is the Filament.

Fig. 8.

In the Valve, the electrode called the grid may be in the form of a cylinder of metal gauze between the filament and the plate, which latter may take the form of an outer metal cylinder (see Fig. 8). If a telephone receiver is connected up in series with a battery across the filament and plate, there will be no sound in the telephone, but if the grid is connected to the receiving circuit so that the oscillations produced in the latter by the incoming signals cause variations of electrical pressure on the grid, then the space between the filament and plate becomes conductive, and currents from the battery flow across this space and produce sounds in the telephone.

The Valve may thus be looked on as a relay, the minute oscillatory currents received being used to release currents from a battery which can work a telephone or other suitable receiver. This relay action may be further utilized by connecting up a number of valves in such a way that the released battery current from one valve is applied to the grid of the next one so as to release a larger current, and so on. When used in this manner to increase the effects of the received currents there is, of course, a practical limit to the amount of magnification that can be produced by increasing the number of valves. It is this magnification effect that makes the Three Electrode Valve such an enormous advance over all other receiving devices, and it is also one of the properties which make it so valuable as a wireless transmitter.

It was by means of this apparatus used as a transmitter that clear wireless telephone messages were sent across the Atlantic for the first time from Arlington, near Washington, to Paris, in 1915, and very early in the war it was used as a receiver by the French to intercept, by earth conduction, messages sent on field telegraph and telephone lines, a practice which both the Germans and ourselves quickly followed. At the time, great surprise was expressed in this country at the intimate knowledge which the enemy possessed of our intended dispositions in the front line, and a similar surprise at our knowledge was no doubt just as prevalent in Germany. It was not long, however, before both sides had adopted means of sending messages along the wires in such a way as to make interception by the opposing side almost impossible, but there can be no doubt that the introduction of the Three Electrode Valve, which has so increased the possibilities of communication, involved at the outset the loss of many gallant lives.

Owing to its magnifying power the Three Electrode Valve can be used for operating high-speed recording devices, and by using automatic high-speed transmitters the rate of signalling can be very greatly increased. In practice, the greatest speed that can be attained by hand transmission and aural reception is about 30 words a minute, and a steady 20 words a minute is considered to be a good working speed. By automatic transmission and reception, using the Three Electrode Valve, a working speed of 200 words a minute is even now quite practicable, and much higher speeds have been experimentally obtained.

It must not be thought that these very high speeds are yet a practical proposition for continuous working day in and day out throughout the year, especially over long distances, owing principally to the trouble caused by atmospherics, which, as previously mentioned, is the really great hindrance to the efficiency of wireless signalling. Over short distances, however, in temperate climates continuous high speed working is becoming quite a normal arrangement. When signals are being received at high speeds, that is, speeds over 30 words a minute, they cannot be read by ear, so that some sort of recording apparatus must be used. Until the introduction of the Three Electrode Valve as an amplifier of currents in the receiving circuit, there was always difficulty in obtaining currents in the receiver of sufficient strength to work a relay at high speed, and so bring into action a recording device. A relay is an instrument by which a relatively small current can bring into action a much larger current, which can be utilized to work, for example, a recording device, just as the slight pressure on the trigger of a gun can bring into action a great explosive force. By this means the extremely weak currents produced in the receiving antenna by the waves from the sending station, can be made to actuate the robust high speed recording devices used in ordinary line telegraphy. In some of these devices the message is recorded in the Morse code on a paper tape, and is read off and written up by an operator, or it may appear as punched slip, similar to the punched slip used at the transmitting station, and be then run through an instrument which produces the message in printed characters on another slip. In fact the Valve, by its magnifying properties, has thrown open to wireless the high-speed recording devices which have been developed after years of trial on line telegraph and telephone circuits.

# CHAPTER V

## WIRELESS TELEPHONY

IN communicating by speech in the ordinary way, energy is transferred from the speaker to the listener in the form of air waves, these waves being produced by oscillations, to and fro, of the little particles of air. These air oscillations are set up by the voice, and are received as air oscillations in the ear, where they act on the nerves and produce the sensation of sound. In line telephony the air oscillations set up by the voice actuate an instrument, the microphone transmitter, which varies the strength of the electric current flowing along the line. These variations of the current work the receiving instrument, the telephone receiver, in such a way as to produce sound waves precisely similar to those which caused the variations at the sending end. In this way the energy of the air oscillations at the transmitter is transformed into the energy of current variations in the line, and re-transformed into sound at the receiver.

In the case of wireless telephony the procedure is very similar, except that the variations of a current flowing in the system of elevated wires, called the antenna, produce variations in the continuous stream of ether waves which this current causes to be radiated in all directions, and these ether waves cause a current with similar variations to flow in any other antenna which they meet.

In ordinary speech, therefore, air waves proceed direct from the sender to the receiver; in line telephony they vary an electric current which produces similar air waves at the receiver; and in wireless telephony they vary an electric current which produces variations in ether waves, which, in their turn, produce air waves at the receiver.

The medium by which energy is transferred in line telephony is the wire connecting the sender to the receiver, whereas in wireless telephony the medium is the ether which connects all senders to all receivers, and it is this fact which constitutes the great advantage of the latter, viz., all senders are at no cost automatically connected to all receivers. It must not be thought, however, that such a great advantage can be attained without dragging some serious disadvantages in its train. The use of the same medium, for instance, leads naturally to difficulties in selectivity, that is, one communication being interfered with by others, or by electrical disturbances in the ether itself—i.e., atmospherics. Interference between various communications is overcome to a great extent by arranging that different lengths of waves in the ether are used for different communications, the apparatus for one communication being adjusted to respond much better to ether waves of one special length than to those of any other length. When the apparatus at each end are thus arranged for transmitting and receiving respectively ether waves of the same length, they are said to be in tune with one another, and the greater the number of different communications required, the greater are the number of wave-lengths that must be arranged for. Now it is just this requirement that makes an extensive network of different communications very difficult and often impracticable, because for good working over approximately the same range, the various wave-lengths which can be economically used are strictly limited. Wireless telegraphy makes use also of ether waves, so that the available field for selection is still further reduced. With line telephony, on the other hand, there is, of course, no such limitation, separate channels being used for each of the various communications.

The other trouble mentioned, that of interference from electrical disturbances in the ether, is also very serious, just as it is in the case of wireless telegraphy. These disturbances, or "atmospherics," are very troublesome, indeed, in tropical climates, and even in temperate climates they produce much interference in the summer months. For "broadcasting" a message—that is, sending a message out to all stations within range which wish to receive it—the difficulty as regards wave-lengths is of course reduced, and it is in this direction that the first great strides are being made in the use of wireless telephony. The commercial possibilities of broadcasting by wireless telephony were recognized immediately after the close of the war, and in 1920 many promising demonstrations took place. One of the most successful was given from the Marconi Company's station at Chelmsford, where Dame Melba sang songs which were heard clearly at many wireless stations on land and sea up to distances of several hundreds of miles. In the same year the Press Association carried out tests, under the supervision of the Post Office, to see how far wireless telephony could be used for the broadcasting of news messages throughout the country; but the results then obtained did not justify further action on those lines. One point is that the news can be read by anyone with a receiver within the range of action of the broadcasting station, and agencies and newspapers naturally do not wish to have their news received simultaneously by all and sundry free of charge. However, there is no doubt that there, are possibilities for the broadcasting of news by wireless telephony, and that developments may be confidently expected. It was arranged recently that British wireless manufacturing companies could broadcast music and other matters for entertainment purposes on waves between 350 and 420 metres, from eight centres in this country, between 5 p.m. and midnight on weekdays, and at any time on Sundays. This has given a great impetus to popularizing wireless and making its uses, as well as its limitations, more generally appreciated. This popularizing means increased trade for the manufacturer and, retailer of apparatus, and increased interest in wireless possibilities generally throughout the country, both matters of advantage, as can be seen by the results in the somewhat similar case, though on a much larger scale, of the popularizing of motor transport. The advantages of limitations being more generally appreciated will be obvious to any who read wireless "news" in the daily Press.

These limitations, principally that of interference, necessitate a certain amount of Government control to ensure immunity from interference for Government and commercial communications. This fact has always been recognized here, where, indeed, the confined nature of areas for signalling makes it obvious, and has recently had to be recognized in even such a large country as the United States of America, where at first no control was imposed.

At first sight, it would seem that wireless telephony has many advantages over wireless telegraphy, but some of these are more apparent than real. Telephony does not require a telegraphist, but then it does require someone with technical

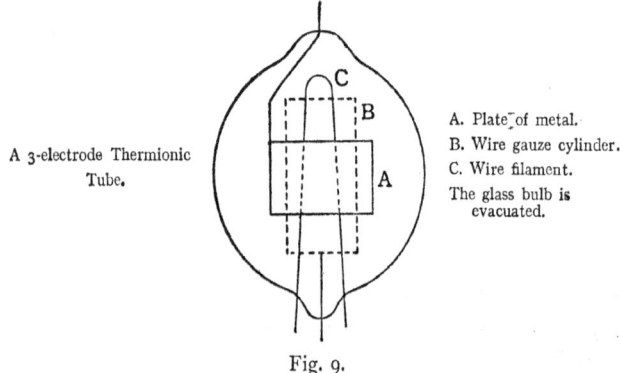

A 3-electrode Thermionic Tube.

A. Plate of metal.
B. Wire gauze cylinder.
C. Wire filament.
The glass bulb is evacuated.

Fig. 9.

knowledge to adjust and maintain the apparatus. The speed of telephony may be greater, but then the efficiency of high-speed telegraphy is rapidly increasing. On the other hand, the electrical power required for communication over any range is much greater for telephony than telegraphy, and the mutual interference between telephonic communications is much greater than between telegraphic communications. It is obvious that the ideal for mobile telephone communications is to connect up ships and aircraft by wireless telephony to land telephone services, and although this has been done already on an experimental basis, it is far yet from being a working commercial proposition.

In comparison with wireless telegraphy, the advances in

Carbon electrode from an arc transmitter at the
Oxford Station.

Base of a steel mast at the Oxford Station.

[*To face page* 52.

wireless telephony were extremely slow until about 1915, when the Three Electrode Thermionic Tube was introduced as a generator of wireless waves. Since then, however, such rapid progress has been made that its great future is already well assured. The principal reason why telephony lagged so far behind was that a continuous radiation of uniform ether waves which could be modulated by the voice was essential, whereas in telegraphy a radiation of groups of waves sufficed, a radiation such as was obtained from the spark generators used in all the-earlier wireless telegraph transmitters. The first real attempts at wireless telephony were made possible when arc generators and high frequency alternators were introduced. Even these, however, did not fulfil all conditions. and it was undoubtedly the introduction of the Three Electrode Tube as a generator that gave wireless telephony a fair start. This Tube, or, as it is often called, Valve, consists essentially, as explained before, of a vacuum bulb similar to the ordinary incandescent lamp, but with two metallic electrodes inside the bulb in addition to the filament (see Fig. 9). When this instrument is connected with suitable circuits to a source of electric power, electric oscillations— that is, electric currents oscillating at very high speeds— are generated, and when these currents are caused to flow in an antenna, a continuous stream of uniform ether waves is radiated. For telegraphy, these oscillations, and hence the waves, are modulated in short and long groups by a signalling key, corresponding to the dots and dashes of the Morse code; but for telephony they are modulated by the voice speaking into a microphone, just as the current in the line is modulated in line telephony. A common form of microphone consists of a small chamber filled with carbon granules through which the current flows, the current remaining constant so long as the granules are quiescent, but varying in strength, owing to the varying electrical resistance, as the granules are made to vibrate by the voice.

The microphone transmitter may be placed as at M (see Fig. 10) in the antenna itself, so as to modulate the currents therein, or may be placed in the Valve circuits, so as to modulate the currents at the source, the energy in either case being transferred from the Valve circuits to the antenna by the currents in coil A inducing exactly similar currents in coil B. At the receiving end, the currents produced in the aerial by the wireless waves are exactly similar in form to those in the sender's aerial and flow through coil C, inducing exactly

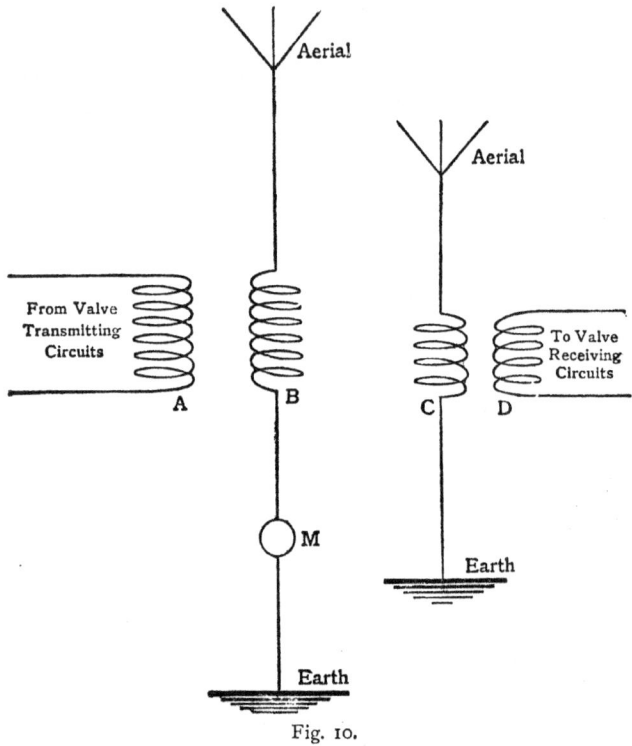

Fig. 10.

similar currents in coil D, which is connected to the Valve receiving circuits in which the telephone receiver is inserted. The aerial in land and ship stations is supported by masts (see Fig. 11), and in aircraft is let down as a trailing wire, or made up as a coil in the aircraft itself. The "earth," consists of metal plates or wires in the ground or water, or as in the case of aircraft, metallic portions of the body.

It was only a year or so before the outbreak of war that experiments were commenced with wireless telephony, using the Three Electrode Valve as a transmitter, and during the war nearly all work carried out in Europe on wireless telephony was directed towards the development of small sets, especially in connection with aircraft. This was largely due to the fact that in aeroplanes the use of wireless telegraphy entailed the pilot or observer being trained as a telegraphist and acting as such in addition to his other duties, as it was usually impracticable to carry a wireless operator for wireless duties only. Pilots and observers had plenty of other work to get through, and the advantage of having a simple telephone set which could be used with little technical knowledge soon became very apparent. It was in the summer of 1915 in this country that wireless speech was first received from an aeroplane, and there is no doubt that our Air Force was the pioneer in this method of aircraft communication. One of the first successful demonstrations of wireless telephone working from aircraft in France was given to the late Lord Kitchener at St. Omer. Unfortunately, however, the messages from the aeroplane were heard as far away as Lowestoft, and further experiments were prohibited for quite a long time, as it was thought that the enemy might obtain some valuable information by reading our messages.

In 1915 most pilots considered all wireless more as a nuisance than anything else, whereas a few years later some pilots on the postal routes refused to go up without a wireless telephone. About the time of the Armistice aeroplanes were normally communicating by wireless telephony with ground stations up to about fifty miles, and with each other up to about five miles, but these ranges have now been more than doubled.

In the United States, during the early part of the war, more time could be spared for experiments with larger sets, and spoken messages from Washington were heard at the Eiffel Tower Station in Paris, about 2,300 miles, and sometimes even at Honolulu, about 5,000 miles. Since then other great distances have been accomplished, such as    between this country and America, but it must be understood that such

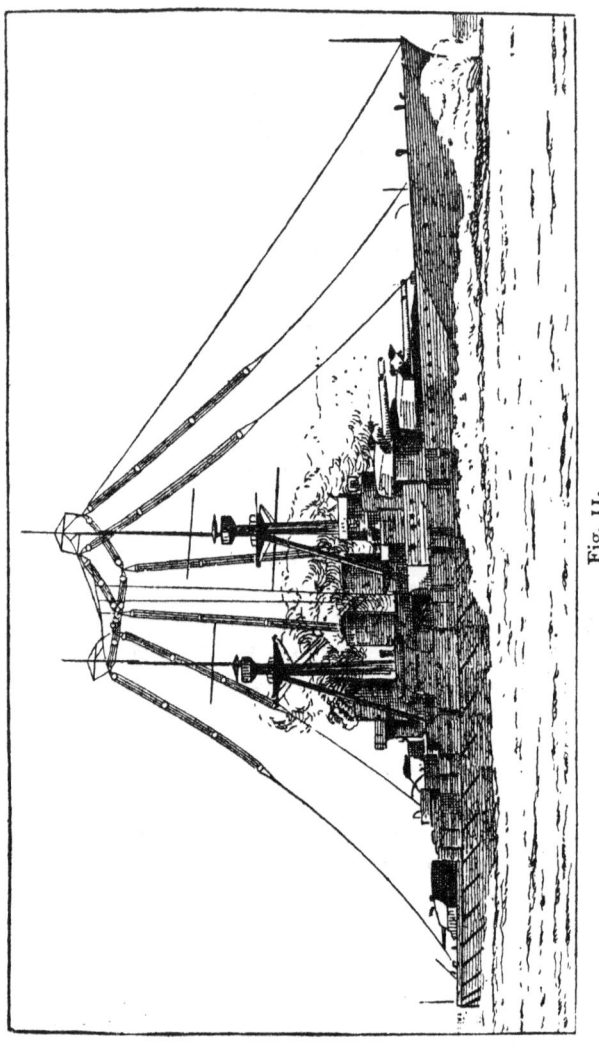

Fig. 11.

long-range working is not yet on a commercial basis.

There is no great technical difficulty in linking up wireless telephony with line telephony in special cases to allow of the communication being partly by wireless and partly by line. One of the first really successful demonstrations of combined wireless and line working was given when President Wilson, when at sea returning to the United States after the Peace Conference, received telephone messages from his Ministers at Washington. At that time one of the great drawbacks to the wireless telephone was that the receiver could not break in, as he can on a land line, when the sender was speaking. Arrangements have now been devised for getting over this trouble, and the Marconi Company gave a demonstration in 1921 of combined wire and wireless telephony between London and Amsterdam, in which the method of working at each end was the same as would have been the case in line working. The first commercial combined wireless and line telephone service was opened in 1920 between California and Avalon in Santa Catalina Island, a distance of about thirty miles. Messages are sent sometimes by this method to aircraft on the London-Paris route, but no commercial service has yet been arranged.

Besides the inherent advantage of wireless telephony for broadcasting, that of a transmitter being connected to all receivers within range, it has another great advantage for long-range working, which is that in wireless there is very little voice distortion. This is, of course, a vital advantage for long-range working compared with submarine telephony, and opens up to wireless telephony a vast field of utility for communication across oceans; but we need to be patient in this matter, as many technical difficulties must be overcome before we can look for commercial success. Enough, however, has been said to show that when consideration is given to the present achievements of wireless telephony, in light of the fact that the whole subject is in its infancy, it is easy to foresee great possibilities, even if we cannot reach those heights of technical imagination so easily scaled by the daily Press.

# CHAPTER VI

## WIRELESS IN SHIPS

The first, and still by far the most important, application of wireless signalling was for communication with ships; in fact, even yet there are no regulations governing international wireless working, except those contained in the International Radiotelegraph Convention of 1912, which deals only with ships' wireless telegraphic communications.

The Conference for the succeeding Convention should have taken place in 1917, but was postponed owing to the war. It is hoped that it will be arranged for 1923, and preliminary conferences amongst the principal allied and associated Powers have already taken place. The scope of wireless communication has extended so much in the last ten years that the new Convention must necessarily lay down regulations for point-to-point and aircraft communications, including wireless telephony, as well as revise those for ships.

So far as ship communications are concerned, the wireless telephone is in no greater use to-day than it was eight years ago, but in the case of aircraft it forms the principal means of communication with the ground and between the aircraft themselves. This is largely due to the fact, as mentioned before, that most aircraft can ill afford to carry a trained telegraphist to do the work which can be done, though on a more limited scale, by the pilot himself making use of a wireless telephone.

Before long, no doubt, the telephone will come into more extensive use for ships, but for some time to come it is likely to remain as much a luxury for ships as the telegraph is for aircraft. The advantages of the telegraph over the telephone are that much longer ranges can be obtained for the same expenditure of electrical energy, and that being more selective, it is less liable to be "jammed;" the disadvantages are that the communication is slower, and that a telegraphist is required to work the apparatus, though it must be remembered that the maintenance, and even the operation, of a wireless telephone requires a certain amount of training. There is an obvious future for telephony in ships when arrangements can be made for communication through land stations direct with subscribers on the telephone system of the country; but we had better confine ourselves to the present time, when, as a matter of fact, such arrangements are not in operation, and wireless telephony is not used for ship and shore communications. On the other hand, wireless telegraphy is now universally recognized as an essential fitting in ships for the safety of life at sea, and in this country has been made compulsory for all sea-going ships which are passenger steamers, or ships of 1,600 tons gross tonnage or upwards. This includes over 3,000 British ships. In vessels carrying 200 or more persons, three operators are normally carried, and a continuous watch is kept. In those with 50 to 200 persons continuous watch is also kept, one operator and at least two "watchers" being carried, and in the remaining ships, watch is kept for eight hours daily at specified times, one operator only being carried. The qualifications required in the case of a watcher are only such as to ensure his recognizing the general distress call and the safety signal. Should he hear either of these calls, he turns over the watch to a skilled operator. Any person in the ship, if so qualified, may act as a watcher.

In this country the Post Office issues licences to ships in respect of their installations, and the certificates for operators and watchers to candidates who qualify at examinations conducted by that department.

The qualifications required for operators were laid down in the 1912 Convention, and require a knowledge of the working and adjustment of the apparatus, a knowledge of the regulations applying to the exchange of communications, and manipulative ability to send and receive messages at a speed of not less than twenty words a minute.

These men are trained at some forty private schools

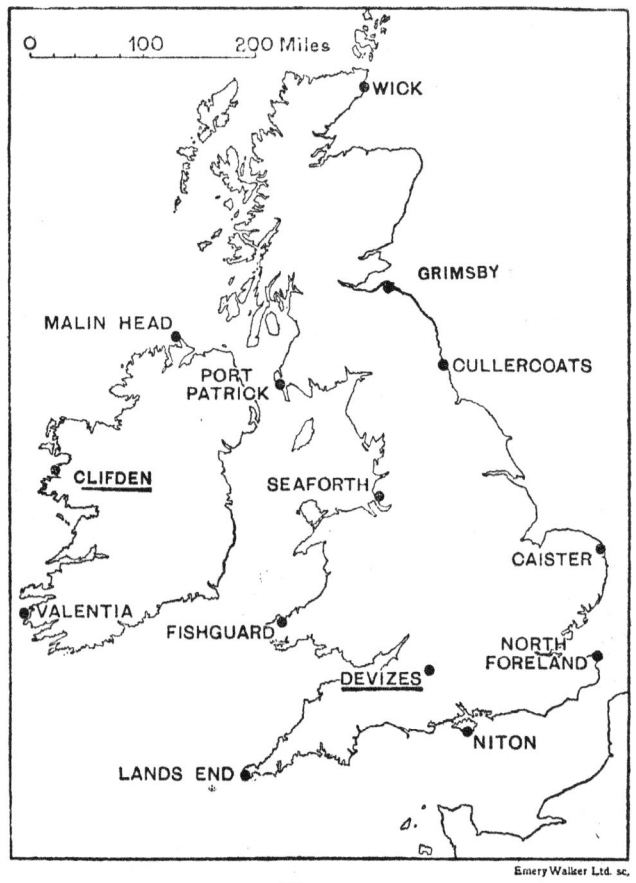

Fig. 12.

distributed over the kingdom, and there are at present about four thousand employed. Reports of irregular working by operators are forwarded by ships and shore stations to the Post Office, where the matters are investigated and appropriate action taken.

It will be understood that any irregular operating in wireless telegraphy is far more serious than in line working, as the interference and waste of time caused by such irregularities are very great, owing to the fact that nearly all signalling is carried out on the same wave-length, viz., 600 metres.

The land stations for communication with ships (see Fig. 12) are all worked by the Post Office, with the exception of the station at Clifden, which is used partly for broadcasting Press and other messages to ships—that is, sending out messages without receiving replies. This station is worked by the Marconi Company. Similar long-range transmission to ships is also carried out by the Post Office station at Oxford. The station at Devizes is fitted with continuous wave apparatus, and communicates up to distances of about 1,200 miles on a special wave-length of 2,100 metres with such of the large liners as are similarly fitted, in addition to their compulsory spark installations. The other stations use spark apparatus similar to that compulsorily fitted in ships, and keep watch on the 600 metres wave, which is that used for distress messages, and, in fact, for practically all messages sent by ships with the exception of the long-distance traffic mentioned above. The compulsorily fitted normal installation in ships must be capable of communicating with a standard Post Office station by day up to at least 150 miles; but communication up to 200 or 300 miles, is usually obtained according to the power of the installation fitted.

An emergency installation must also be fitted, unless the normal one complies with the emergency requirements. These requirements are to the effect that the installation can be worked from a source of energy independent of the ship's steam and electricity supply systems for at least six hours over ranges of at least 120, or 75, miles from a standard Post Office station, according as to whether the ship carries 200 (or more) persons, or carries less than 200 persons. The idea of this emergency installation is that the ship will be able to communicate by wireless in emergencies, even if the normal electrical power supply is out of action.

Ships throughout the world which were fitted with wireless before it became compulsory in this country, were fitted with Spark systems, as Continuous Wave systems had not then been sufficiently developed; so it was laid down that compulsorily fitted installations should consist of Spark sets, as these would be suitable for communicating with all ships; and with those land stations which are used for ship and shore communications.

As a general rule, ships keep watch with their receiving apparatus adjusted for the reception of the 6oo-metres wave, and are thus ready to receive any message of distress from a ship within range, or a call from a ship or land station. The wavelength of 600 metres was detailed for this purpose in the 1912 Convention, as it is a suitable wave for transmission from ships of medium size, and it was very necessary to lay down some one definite wave on which distress messages would be sent and would have as good a chance as possible of being received by all ships and land stations within range. A distress message is preceded by a signal consisting of three dots, three dashes, and three dots sent as one sign, and repeated at short intervals. This is usually alluded to as the S O S signal, though the letters S O S sent as a group in the Morse code is not really the same as the distress signal, which is sent as one sign. When this signal is received all communication ceases, except that consequent upon the call for help. The safety signal consists of three dashes repeated ten times at short intervals, and is sent as a preliminary to information of an urgent character involving the safety of navigation, e.g., icebergs, derelicts, cyclones, etc. The idea of adopting some such warning signal resulted from the inquiry into the loss of the liner *Titanic,* which struck an iceberg and sank on the I5th April, 1912. In this case the S O S signal was responsible for saving over 700 lives, and if a safety signal had been in general use, it is quite likely that the whole terrible disaster would never have occurred.

If satisfactory automatic apparatus for registering the distress signal is devised, British ships, now only compelled to carry one operator, will also be fitted with this apparatus, and ships which carry watchers as well as an operator may substitute the apparatus for the watchers.

The S O S signal itself is unfortunately not suitable for actuating automatic apparatus through the interference

Power house, Cairo Station.

[To face page 62.

caused by other signalling; but some form of alarm signal which could immediately precede the S O S and be suitable for the automatic apparatus, will perhaps be adopted internationally at the next Convention, in which case the necessity for "watchers" will disappear.

The international regulations framed at the 1912 Convention for maritime wireless signalling are in much need of revision, as the great strides made in wireless practice during the war have naturally overstepped the somewhat restricted boundaries of the Convention.

As a matter of fact, the apparatus used in merchant ships was much the same at the end of the war as at the beginning. All the great developments had taken place in naval, military and air force stations, both mobile and fixed, and little time was left for improving the design of apparatus for the mercantile marine. As soon as the war was over, however, the technical lessons learnt were quickly applied to the design of ships' sets. Old sets have now been overhauled, and when new ones are installed their all-round efficiency is far in advance of the pre-war type. In the larger passenger steamers a long-range set, using either the Valve or the Arc system, is being fitted, in addition to the Spark set, and altogether the technical arrangements have been brought into line with the improvements made in naval ship sets during the war. Many naval requirements are, of course, not applicable to the mercantile marine; in fact, the technical lines of development in the two services can never run on exactly parallel lines, as a warship is normally required to fit in for communication purposes with a fleet in close touch, several ships of which may be signalling simultaneously on different wave-lengths, whereas a merchant ship is normally acting as a more or less isolated unit, always ready to receive a distress call on the 600-metres wave-length from any ship at extreme range. This means that for a warship very selective apparatus is an absolute essential, and it can be readily arranged, as a navy is a small unit where control is easy; but in merchant ships great selectivity may often be a distinct disadvantage, as calls from ships not accurately adjusted to the wave-length intended may be missed, and it is impossible at present to ensure that the apparatus in all ships of all nations shall be accurately adjusted. Selectivity, like Free Trade, is excellent if everyone else adopts it.

Besides long-distance continuous wave sets, with the possibility of telephony in the background, the important advent of directional apparatus must be noted. The wireless "direction finder," or compass, is becoming of recognized value in navigation when other means fail, as in fog. The apparatus is fitted in the wireless office, and is worked by the wireless operator. With this apparatus the operator can obtain the bearing of a wireless station relative to the fore and aft line of the ship, the position of the ship being obtained by the intersection of bearings from two or more stations, up to about 100 miles the bearings as set off on the Mercator's chart are accurate enough for all practical purposes, but for greater distances conversion tables or charts on the gnomonic projection must be used. In these charts great circles appear as straight lines instead of as curves, as in the case of Mercator's, and the fact that such charts are required for long distances will be evident when it is remembered that wireless waves travel along great circles over the surface of the globe. Usually the bearings obtained are correct within two degrees, and greater accuracy is not often required, but errors, mostly due to the bending effects experienced by wireless waves, especially at night, do often give trouble. A number of ships have been fitted with this direction-finding apparatus, and have often proved its utility as an aid to navigation in thick weather. There are some Government stations designed and used solely for the purpose of giving bearings to ships. In these cases the station or stations take bearings by signals transmitted from the ship, and communicate them to the ship. The results are, as explained in the next chapter, more accurate than those obtained by having the directional apparatus in the ship, but the advantage of the ship itself being able to obtain its bearing from any ordinary station within range is obvious. Another advantage of having the apparatus on board is that in foggy weather it may be used for obtaining the direction of another ship fitted with wireless.

Advances are also being made with directional transmission—that is, the transmission of wireless waves in one direction only, instead of the usual all-round radiation, and the uses of such an arrangement for navigational purposes are obvious.

The charges for messages to and from ships are at the rate normally of 11d. a word, i.e., 6d. for the station, 4d. for the ship, and 1d. for the inland telegraph. Messages for ships can be handed in at any Post Office, whence they are telegraphed direct or through the General Post Office, London, to the appropriate land station, the procedure being for a ship to notify the nearest land station of its position, course, speed, and destination, so as to facilitate this routing of telegrams. In some cases, as for cross-Channel boats, the charges are reduced, and in the case of the long-distance telegrams sent from Oxford and Clifden the total charge is raised to eighteen-pence a word. These high-power stations, Oxford (Post Office, England) and Clifden (Marconi Company, Ireland), are really point-to-point stations, Oxford normally working to a similar Post Office station at Cairo, and Clifden being used for trans-Atlantic communications. In addition, the Post Office works stations at Northolt (London), Caister (Norfolk) and Stonehaven (Kincardineshire) for Continental communications, and the Marconi Company stations at Carnarvon (Wales), Ongar (Essex) and Chelmsford (Essex) for American and Continental communications.

# CHAPTER VII

## DIRECTIONAL WIRELESS

A S a rule, at small wireless stations, such as those for communication with ships and those in the ships themselves, no attempt is made to obtain any directive effects, that is to say, the waves from the sending station are radiated in all directions, and at the receiving station waves can be received equally well from all directions.

For the ordinary purposes of communication between ships, and between ships and shore stations, it is indeed essential that all-round signalling should be the normal arrangement, so that a distress signal from a ship will be heard by all ships and stations within range, irrespective of direction; but it may often happen, especially in the case of ships and aircraft, that an indication of the direction from which signals are being received will be of great value. This fact was grasped very early in the war by the Air Forces on both sides, and directional apparatus for navigational purposes was rapidly developed and extensively used in military aircraft. For ships, on the other hand, the development ran on another line, that of locating enemy ships, which were using wireless, by means of direction-finding shore stations, and stations used for this purpose by both sides round the North Sea are shown in Fig. 13. These stations, at any rate those on our side, did excellent work in obtaining timely information on many occasions of the movements of enemy craft, both ships (including submarines) and aircraft. The most famous occasion was that leading up to the Battle of Jutland. This was well told at a meeting of the Institution of Electrical Engineers by Admiral of the Fleet Sir Henry Jackson in the following words:

"We have heard much about the use of direction-finding for minor tactical movements of all arms, but this is a case of a major strategical operation which brought about the historical meeting of the British and German Fleets at the Battle of Jutland on May 31, 1916.

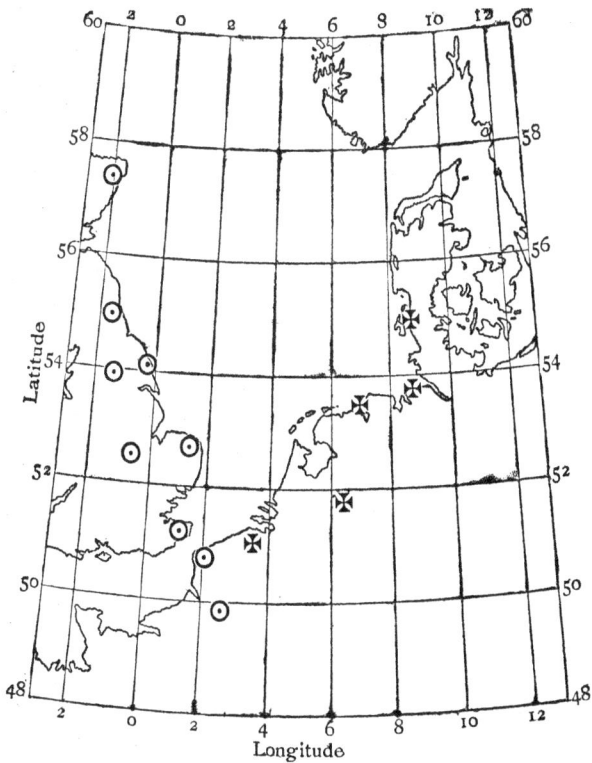

Fig. 13.—BRITISH AND GERMAN DIRECTIONAL WIRELESS STATIONS.

I was First Sea Lord at the time, and so was responsible for the disposition of the Grand Fleet. I may incidentally mention that, in spite of other statements of which I have heard, its Commander-in-Chief (Lord Jellicoe) and I lived, so to speak, with the object of bringing off such a meeting. Our wireless direction-finding stations, under Captain Round, kept careful and very intelligent watch on the positions of German ships using wireless, and on May 30, 1916, heard an unusual amount of wireless signals from one of the enemy ships which they located at Wilhelmshaven. This was reported to me; the time was a critical and anxious one in the war, and I had also some reasons for expecting that the German Fleet might put out to sea during the week. Our Fleet was ready at short notice and had arranged, unless otherwise prevented, to put to sea on the following day for a sweep of the North Sea. But if the German Fleet got to sea first, the chance of a meeting in waters not unfavourable to us was remote; our object was to try to get to sea before or shortly after the Germans, and hitherto we had not succeeded in doing so. Later on in the afternoon, it was reported to me that the German ship conducting the wireless had changed her position a few miles to the northward. Evidently she and her consorts had left the basins at Wilhelmshaven and had taken up a position in the Jade River ready to put to sea. This movement decided me to send our Grand Fleet to sea, and move towards the German Bight at once and try to meet the German Fleet and bring it to action. This they did with their usual promptitude, and the result was the famous Battle of Jutland, and it was indirectly brought about by the careful and accurate work of Captain Round and his staff, for which I hope they will now accept my belated thanks and appreciation. Their work is not ended. Direction-finding has come to stay for more general use in peace. Errors are being eliminated, and there should be a great future before it, especially on the lines indicated in the Press to-day by the Admiralty for assisting navigation at sea as well as in the air."

It will be gathered from Admiral Jackson's closing remarks that some of the war-time direction-finding stations are still being used, but for the more peaceful purpose of assisting in the navigation of ships and aircraft. The stations were not used for that purpose to any great extent in the war, though directional sets fitted on aircraft were used pretty frequently for navigational purposes by obtaining bearings from known fixed land stations.

At first sight, it would appear that directional receiving sets in ships or aircraft would be invaluable for obtaining bearings from land stations, and so they would be if the results obtained were always reliable. There are, however, errors which are liable to arise from various causes, some of which are not yet fully understood, and cannot therefore be allowed for. The most serious of these arise from bendings of the paths of the waves by reflection and refraction due to certain states of the atmosphere; for instance, observations may be made at, say, 100 miles, within an accuracy of 1 degree on nineteen days out of twenty, but on the twentieth day there may be an error of as much as, say, 3 degrees, due to some bending effect on the path of the waves taking place between the sending and receiving stations. There are also possibilities of local errors due to the effects of metal-work, such as stays for masts, etc., at the receiving station, as well as bending effects on the waves due to the proximity of cliffs, or long coast-lines in the path of the waves. These local errors, since their causes are better understood, can be eliminated, or corrected, sufficiently for all practical purposes, but it is clear that, at the best, the bearings given by wireless directional methods cannot yet be looked on as thoroughly reliable. The point is, however, that they are sufficiently reliable to be very useful when it is impossible to take observations by other means, due to, say, foggy weather.

We might now examine briefly the principles underlying directional wireless, and in doing so let us first recall the usual non-directional arrangement. At the transmitting station, wireless waves are radiated in all directions by oscillatory electric currents flowing in an elevated system of wires, called the aerial or antenna, which in its simplest form, and that first used, consists of a single vertical wire. At the receiving station these waves produce similar, though much weaker, currents in the receiving antenna, which in its simplest form consists also of a single vertical wire, and there is no directional effect; that is to say, the direction from which the waves are coming has no effect on the strength of the signals received. As soon, however, as the antenna is bent or sloped out of the vertical, so does it become better for sending to, or for receiving from, one direction more than any other This effect was foreshadowed in Hertz's original experiments in Germany in 1887, but was not made much practical use of until 1905, when Marconi commenced to use an inverted L-shaped antenna for transmission and reception. Such antennae transmit best to, and receive best from, the direction opposite to that towards which the horizontal portion stretches. Antennae of this description are used at many high-power stations, so that the greatest possible radiation is directed towards the receiving stations.

The writer attended the first demonstration given in 1905 with this arrangement by Senatore Marconi to Government officials. It took place after dinner in the hotel at Poldhu a few hundred yards from the high-power wireless station. Mr. Marconi placed a magnetic detector on the table with one end of an antenna, which consisted of a couple of feet of

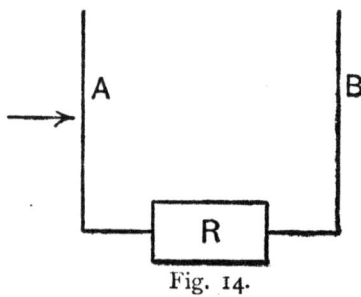

Fig. 14.

wire, joined up to the instrument in the usual way. He walked round the table with the other end, and when the wire was in the direction of the station, loud signals could be heard in the telephone receivers connected to the magnetic detector; when at right angles to this direction no signals could be heard. It was a simple and impressive demonstration of directive effects, then quite new and surprising.

Another form of directional antenna which is very extensively used, especially for receiving stations, consists of a vertical rectangular loop which best receives waves travelling in the direction of the horizontal sides. Signals from any other direction are weaker, until in the line at right angles to the best direction no signals are heard.

This effect can be easily seen from a consideration of Fig. 14, where A and B are two vertical wires, and R is the receiving apparatus. If the waves are coming from the direction shown by the arrow, or in the exactly opposite direction, their effect on A will not be the same, at the same moment, as their effect on B, and some current will flow through the receiver R. If, however, the waves are coming in a direction at right angles to this, i.e., through the paper, they do produce the same effects in A and B at the same moment, and no effect is produced in R. This is clear if one considers the moment when the waves produce currents of the same strength flowing down A and B simultaneously, as then, so far as R is concerned, these currents

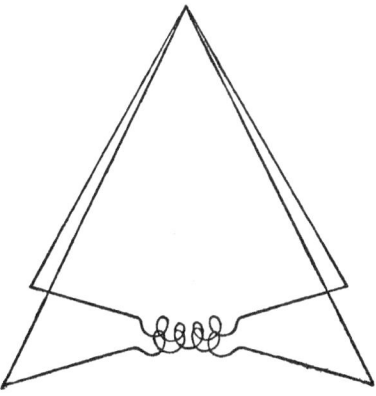

Fig. 15.—TRIANGULAR LOOP AERIALS.

cancel one another and produce no effect. This arrangement of directional aerials was first introduced by S. G. Brown in this country in 1899. The Italians, Bellini and Tosi, used triangular loop aerials, as in Fig. 15, at right angles to one another, in conjunction with a special receiving circuit, in which a coil is revolved by hand until signals become weaker and weaker and eventually disappear, when a pointer attached to the coil indicates on a scale the direction from which the signals are being sent. The region of maximum signals instead of minimum is utilized in the latest forms of this apparatus, but these are technical details which need not be considered here.

This Bellini-Tosi arrangement was   used for several years before the war, but was not fully developed until, during the war, the introduction of the three-electrode thermionic tube as an amplifier of weak signals completely revolutionized wireless reception, and made it possible to employ small loop aerials for the reception of signals transmitted from even great distances. Small loops are now extensively used for direction-finding purposes in aircraft, and larger ones in ships. The adoption of this apparatus in ships is not so general as in aircraft, but a few score of our merchant ships are already fitted. The possible errors already mentioned largely account for this comparatively slow growth, as well as the fact that directional shore stations can give more accurate bearings to ships than ships can obtain with their own directional apparatus.

In France, nine of these shore stations are already in operation along the English Channel alone, and in the United States about a hundred are contemplated, of which more than half are working. In this country there are only four in operation. In shore stations greater accuracy is obtained than in ships, as there are less obstructions to the path of the waves in the form of masts, stays, etc., and more highly trained operators, as well as conveniences in the matter of space, quietness, and stability, are available.

A ship or aircraft fitted with directional apparatus can obtain a bearing relative to its fore and aft line from any wireless station which it hears transmitting, and by taking such bearings from two, or preferably three, stations, it can obtain its position, usually to a degree of accuracy sufficient for the purposes of navigation when other means have failed, provided that the stations happen to be suitably placed for obtaining a good cut with the lines of bearing. For distances up to about 100 miles the bearings can be plotted on the usual Mercator's chart, but for greater distances a chart on the gnomonic projection must be used, as explained in the last chapter. As a matter of fact, in ship work, bearings from stations at distances of over 100 miles are very seldom required, so that the usual Mercator chart gives sufficiently accurate results under favourable conditions; but, as mentioned above, the results obtained by ship sets are not so reliable as those obtained by direction-finding shore stations.

The Northolt Station.

[*To face page* 72.

In the latter case, the ship, which does not require any directional set of its own, calls up a station and asks for its bearing. The ship then transmits a succession of letters, the station takes the bearing and signals it to the ship. Sometimes two or more stations work in conjunction; each takes a bearing, and the controlling station then signals to the ship the position which is obtained from the bearings taken. It will be noticed that when shore stations do the work, the responsibility for the results rests with them, not with the ship, as would be a more satisfactory arrangement if the results obtained by ships were as reliable as those obtained by shore stations. In this country, a charge of five shillings is made for a bearing given by a direction-finding shore station, and similar charges are in contemplation, or are already made, in several other countries.

Directional receiving stations are now coming into extensive use for another, quite different, purpose, viz., commercial long-range point-to-point communication. Three conditions are required for a paying commercial long-range service; first, high-speed signalling; secondly, duplex working; thirdly, freedom from interference, and of these the second and third can be obtained, to a great extent, by receiving the messages at a directional receiving station separated from its sending station. This arrangement is now made for practically all long-range services throughout the world. In duplex working, messages are sent and received at the same' time, the receiving station being some miles distant from its sending station, and arranged directively, so that it can receive from the distant station without being interfered with by the transmission of messages from its own sending station. Similarly, freedom from interference from other stations is materially assisted by the fact that the receiving station, being directional, can only receive well on the line between it and the distant station from which it wishes to receive, so that sending stations situated in other directions will not interfere. Owing to the great amplification in the strength of the current at receiving stations made possible by the use of thermionic tubes, quite small loops consisting of a number of turns of wire wound on rectangular frames, about eight feet square, are now the only aerials used at many of the high-power stations in Europe for receiving commercial messages from stations at great distances such as those in the United States of America.

So far, we have considered directional receiving apparatus only, but transmitting stations can also be arranged to produce stronger radiation in one direction than in any other. The French Government have a station of this sort at Boulogne on the Bellini-Tosi system, but this is the only true directional transmitting station in operation.

An important paper, however, on Short-Wave Directional Wireless Telegraphy, read before the Institution of Electrical Engineers on the 25th of April, 1922, by Mr. C. S. Franklin, of the Marconi Company, opens up what may well be a new era in wireless signalling. Mr. Franklin pointed out that directional wireless was employed by Hertz, who made use of reflectors at both the sending and receiving stations in order to augment the effects, and to prove that the waves obeyed, to a considerable degree, the ordinary optical laws of reflection. Senatore Marconi, too, employed reflectors in his early experiments, but the use of reflectors of reasonable dimensions necessitates very short waves of only a few metres in length. Such waves were soon found to be unsuitable for long ranges, and as the demand in the early days was for long-range all-round working with ships, the reflector method of directing waves remained undeveloped. To-day, however, the practical gamut of wave-lengths suitable for long-range working is becoming fully occupied, and the importance of directional arrangements being developed so as to reduce interference is apparent. Marconi never lost sight of the possibilities of his early experiments, and for war purposes recommenced his investigations in Italy in 1916 on the use of reflectors with very short waves of only a few metres in length. Mr. Franklin, who was assisting Senatore Marconi, stated that the only interference experienced on these waves was from the ignition systems in motor-cars and motor-boats, which apparently emit waves up to about forty metres in length. The reflectors used consisted of a number of wires arranged on a cylindrical parabola with the antenna at the focus, and, with the apparatus available, ranges up to six miles were obtained.

The experiments were continued in this country, and in 1921 directive wireless telephone messages were sent from Hendon to Birmingham, a distance of 97 miles, using extremely short waves of the order of 20 metres. These results open up great possibilities for directional work with ships and aircraft, as well as for providing secrecy and reducing interference in the case of ordinary point-to-point communications.

In connection with directional ship work, trials on these lines are being carried out with revolving reflectors. The idea is that a transmitter and reflector revolving will act as a kind of wireless lighthouse which will be able to give ships their positions during fog, when within about ten miles of the danger-point, the stations being situated at such places as are now occupied by fog signals.

In the United States of America and in France, a number of small transmitting stations, called radiophares, are used to transmit continuously in times of fog, each station radiating some distinctive signal in order that ships fitted with wireless compasses may obtain bearings. These stations are found useful for navigational purposes in narrow waters, such as river mouths and channels.

The great use of directional wireless for aircraft is apparent. If the aircraft is not fitted with directional apparatus, it must rely on direction-finding land stations, and, as these are few and far between, the advisability of having a directional set in the aircraft itself is obvious. In the latest development, the aircraft's set is arranged to give the direction of any sending station when the set is adjusted to produce strongest signals from that station.

# CHAPTER VIII

## THE IMPERIAL WIRELESS CHAIN

IMPERIAL wireless may be said to date from October, 1900, when the Marconi Company commenced the erection of the first high-power station in the world, at Poldhu in Cornwall. Unfortunately, the first check in the long list which has followed came within a year, when the masts at Poldhu were wrecked by heavy gales in the autumn of 1901. The year 1902 was spent in making good this loss, and in further improvements, and on the 18th December, Senatore Marconi opened up a fresh page of history by sending a wireless message from the Cape Breton Station in Nova Scotia to King Edward VII. in England. The following year, 1903, saw the commencement of a regular wireless service to ships at sea, and in 1904 the first imperial service was working, though far indeed from regularly, between this country and Canada. In 1905, the Marconi Company commenced the erection of a high-power station at Clifden on the west coast of Ireland, and in 1907 a commercial transatlantic service was opened between this station and Glace Bay in Nova Scotia.

It had thus taken seven years of unflagging hard work and resourcefulness, in face of what then seemed to many quite insurmountable obstacles, for Marconi and his small band of enthusiastic followers to forge the first link in our imperial wireless communications.

During the next four years, on the practical side, the Clifden-Glace Bay service went on improving, but at a rate which gave no sleepless nights to shareholders in cable companies, and on the theoretical side, the idea of an imperial system progressed equally slowly, until it culminated eventually in a decision at the Imperial Conference of 1911 to the effect that an Imperial Wireless Chain of stations, owned by the State, should be erected without delay. In July, 1912, the Postmaster-General entered into an agreement with the Marconi Company for the erection of the Chain. By this agreement the Company were to erect stations in England, Egypt, East Africa, South Africa, India, Singapore, and Hong-Kong, the chain being extended from Singapore to Australia by the erection of a station by the Australian Government at Port Darwin.

The terms of the agreement gave rise to bitter discussions in Parliament and Press, and were considered in the winter of 1912-13 by a Select Committee of the House of Commons, the technical aspect being dealt with by a committee of experts under the chairmanship of Lord Parker. The main principles of the scheme were accepted, and the revised contract between the Postmaster-General and the Company was ratified by Parliament in August, 1913.

A commencement was made with the construction of the English station at Leafield, near Oxford, and of the Egyptian one at Abu Zabal, near Cairo, but the outbreak of war in 1914 resulted in a change of policy, and this scheme of imperial wireless intercommunication was dropped for a less grandiose one of imperial wireless ship and shore communication. This latter scheme consisted of the erection of medium-power stations, primarily for ship work, at Jamaica, Bermuda, St. John's (Newfoundland), Demerara, Aden, Mauritius, Durban, Port Nolloth, Bathurst (Gambia), Seychelles, Ceylon, Singapore, Hong-Kong, as well as more powerful ones at Ascension and Falkland Islands. Jamaica and Bermuda stations were erected by the Admiralty, and the remainder on behalf of the Admiralty by the Marconi Company. These stations, along with a similar one erected by the Admiralty in the Azores (Portuguese territory), proved of great value during the war for the purpose for which they were designed, viz., communications with ships, and occasional point-to-point strategic communications, but they were not powerful enough, nor could they be adapted, for commercial working as an Imperial Chain.

Other countries, however, adopted a different policy, as point-to-point communications, in their cases, were more important than those of ships. The result is that France has now in operation, or in course of erection, high-power stations at Paris, Bordeaux, Nantes, Lyons, in Algiers, Senegal, the Congo, the Soudan, Madagascar, Cochin China, New Caledonia, Martinique and Papeete. Italy has first-class stations at Rome and Coltano, and two useful ones in Somaliland. Germany had commenced a chain, and on the outbreak of war had powerful stations at Nauen, Norddeich and Eilvese in Germany, and in Togoland (Kamina) and South-West Africa (Windhuk). The United States of America erected stations to link up Washington with Europe, California, the Panama Canal zone, and the Philippines, these last being linked up by stations in Honolulu and Guam.

In this country, as soon as the war was over, and the Marconi Company had been awarded in the law courts £590,000 as compensation, etc., from the Government for the abandonment of the original scheme, the desirability of pushing on with some such scheme, again became apparent, especially as the cables were much congested, and the consequent delays were seriously hampering the reorganization of business throughout the Empire. In 1919, the Government decided to go on at once with the erection by the Post Office of the stations near Oxford and Cairo, now in operation, and to appoint a Committee to go into the whole question of Imperial Wireless communications.

The Committee appointed by the Government, the Imperial Wireless Telegraphy Committee, under the chairmanship of Sir Henry Norman, consisted of three independent wireless experts and four Government officials closely connected with the development of wireless and electrical science generally. This committee's report, dated June, 1920 (see Map), was approved by the Government and by the Imperial Conference of 1921, Australia, however, retaining full freedom of action as to the method in which she would co-operate.

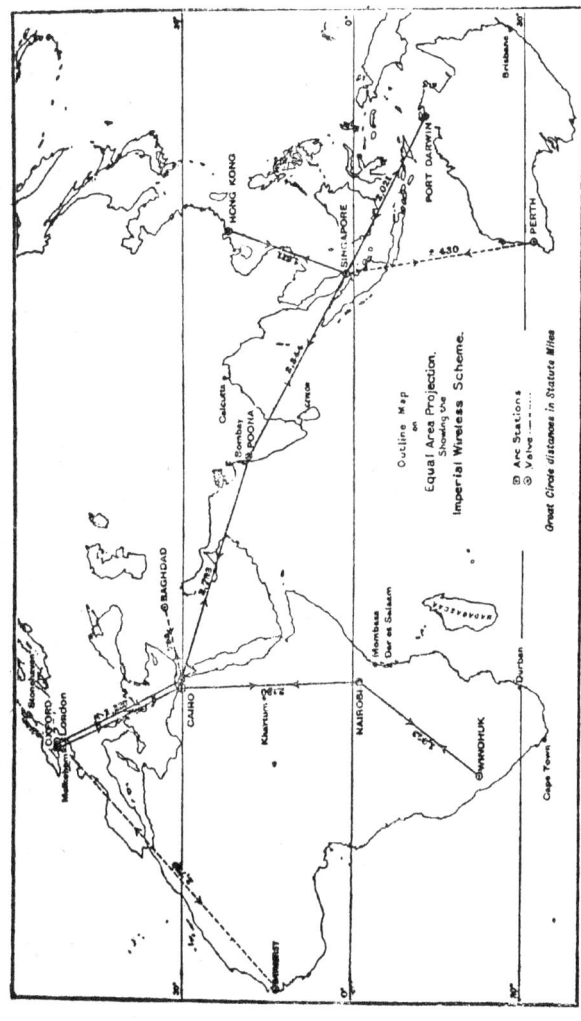

Reproduced from the Report of the Imperial Wireless Telegraphy Committee, 1920.]

The Committee summarized their recommendations and findings as follows:

" We recommend:

"1. That a scheme of Imperial wireless communications be established connecting the communities of the Empire by geographical steps of about 2,000 miles each, as indicated on the accompanying map.

"2. That the wireless system employed be that involving the generation of radio-telegraphic energy by thermionic valves.

"3. That the service of communication between Leafield and Cairo by Foulsen arcs, shortly to be in operation by the Post Office, be the first link in the chain of communication with the British communities in Africa, and that this communication be continued by a valve station near Nairobi, in East Africa, and by the alteration of the ex-German station at Windhuk to a valve station, to complete the connection with the Union of South Africa.

"4. That for communication with India, the Far East, and Australia, valve stations be erected in England, near Cairo, at Poona (or other Indian station), at Singapore, at Hong Kong, and in Australia at Port Darwin or Perth.

"5. That similar communication be established by valve stations between England and Canada, subject to decision in conference between the Imperial and Canadian Governments.

"6. That the stations be planned by a Wireless Commission of about four members, as herein described, whose functions would probably cease with the completion of the stations, and that the construction of the stations be entrusted to the Engineering Department of the General Post Office and the corresponding Dominion and Indian authorities, according to the plans furnished by the Wireless Commission.

" We find:

"1. That an Imperial wireless scheme established in this manner would afford reliable, expeditious and economical communication for commercial, social and Press purposes throughout the Empire, and that it would meet essential Imperial strategic requirements.

"2. That estimates of revenue and expenditure indicate an initial annual loss, after paying interest at 6½ per cent, on capital, and allowing for complete amortization of buildings and plant within a proper period, of about £100,000, divided as shown between the Imperial Government and the other Governments concerned, but that *(a)* this loss, which was to be expected, may reasonably be regarded as likely to decrease annually, until after ten years the service will show a profit; *(b)* the system recommended is probably the most economical that will produce the results required, and at the same time be in accord with present wireless science and future wireless developments; and *(c)* the small temporary loss is negligible in comparison with the Imperial benefits to be conferred."

The Committee estimated that the total capital cost of the scheme, excluding the Canadian link and the existing Oxford-Cairo link, would be £1,243,000, of which £853,000 would fall on the Imperial Government.

The Committee's report gave rise to much criticism on two points—first, that the State is to erect and work the stations; secondly, that instead of having direct working between England and India, between England and Australia, and between England and South Africa, the scheme contemplated the use of intermediate stations for these communications.

As regards erection and ownership by the Government, it was urged that the chain could be erected more rapidly and worked more efficiently by private enterprise than by the State, and the success of such an arrangement in the case of our cable communications was often referred to in this connection. The trouble is, however, that, as was pointed out by the Committee, an Imperial wireless system to work efficiently must be protected from interference from other sources, owing to the very limited number of "wave-lengths" available for long-distance working, and of these "wave-lengths" in any international agreement, only a proportion could be claimed for the British Empire.

In other words, owing to the mutual interference, which is so easily caused by the signalling of high-power wireless stations, any Imperial scheme must be, for all practical purposes, a monopoly, and though there may be objections to a Government monopoly, there are greater objections to a private monopoly. In the case of cables, this monopoly question does not arise, as there is no mutual interference, and therefore no limit to the number of competing cables that may be used for any one connection.

Another important argument for State ownership is that the Crown cannot be prevented from using any patent, and, in fact, in the Committee's opinion, a private monopoly would be deeply prejudicial to research and progress. The Committee, however, were careful to point out that "a State monopoly of this kind would not preclude private enterprise in other spheres of wireless activity, Wireless companies would still have an important sphere as manufacturers, and in providing and operating ship stations, and they might, if desired, be granted concessions on suitable conditions for long-distance communications between British territory and foreign countries."

As regards the superiority of direct working with India, Australia and South Africa, as opposed to utilizing intermediate stations for passing on the messages, there can be no question. The point was, however, could stations then be erected in England, India, Australia and South Africa to give an economical and efficient commercial service at all times? The Committee thought not, and proposed, therefore, the use of intermediate stations for communications between these main centres, as well as for their own communications.

The Wireless Telegraphy Commission recommended by the Committee for planning the stations, was appointed by the Cabinet in December, 1920, under the chairmanship of Lord Milner, and in their report, published in January, 1922, it was made clear that the stations which they were planning for the main centres would be capable of direct working with one another, except during periods of unfavourable atmospheric conditions. For what average periods of the day and night this direct communication will be practicable it is impossible to

The Cairo Station.

[To face page 82.

predict, more than to say that, from experience already obtained at high-power stations, direct working between centres will be possible for a considerable number of hours daily. As mentioned above, Australia retained full freedom as to the method in which she would co-operate, and has now decided to contract with a commercial company for the erection of a station to work direct with this country.

It was announced in July, 1922, that the erection of the new linking stations, as well as the Hong Kong station, would be deferred for the present at any rate, but that the erection of the main stations would be commenced immediately, a site at Bourn, in Lincolnshire, having already been selected for the English station.

# CHAPTER IX

## WIRELESS IN THE HOME

A WAVE of enthusiasm for what the Press aptly calls "wireless in the home" is now passing over the country. In the United States of America this enthusiasm has been in progress for the last year or so, and has now become so universal as to be called "radio flu," a term which shows that its dangerous propensities must be taken into consideration. Here we are apt to move slower in exploiting the developments of science, and in this particular case there is much to be said for —"And look before you ere you leap : For as you sow, y'are like to reap"

In the United States wirelesss signalling, both telegraphy and telephony, had been given almost a free hand, so much so that the Government has been forced to consider means of preventing the almost chaotic conditions which prevailed. But it must not be thought that this free hand method, as a start, was necessarily a wrong one. It introduced the capabilities of wireless to literally millions of people in a striking manner, and at the same time it brought home to these people the limitations of wireless in a way that certainly no other method could have accomplished. It is for us now to realize the facts—that is to say, the capabilities, and the limitations, and to see that our methods of advance take cognizance of both.

Before we can consider the future possibilities we must, or at any rate we should, understand the present facts. It is easy, far too easy, to rush off into the regions of guesswork and generality without having first acquired the solid foundation of knowledge which should be the starting-point for these pleasant excursions into the future.

We must first of all realize that all wireless messages are sent by waves in the hypothetical all-pervading medium, called the ether, and that these waves are usually radiated in all directions, the strength of the received signals decreasing as the distance from the source increases. The question at once arises how can a message be read if another message is being sent at the same time from some station within range: in other words, how can the "jamming" of messages be obviated? If the messages are being sent by waves of the same length, this length being the distance from the crest of one wave to the crest of the next one, the receiver will obtain merely a jumble-up of words or Morse signs, according to whether the message is by telephony or telegraphy. If however, the wave-lengths are different, the receiver can adjust, or as it is called "tune up" his instruments so that they respond only to the wave-length of the message which he wishes to receive. The trouble is that the number of wave-lengths available is limited, and unless some control is exercised chaos must ensue. In line telegraphy and telephony the message is confined to the wire, not as in wireless, thrown out into space. But it is this very difference which makes wireless so suitable for broadcasting—that is, sending messages to everyone within range who wishes to receive them. If the messages are broadcasted in words by telephony so much the better, as few but the professional and the enthusiast can read them if telegraphy is used.

This is the aspect of wireless which has recently taken us by storm, broadcasting by wireless telephony. Much too, of course, is included, everything that the gramophone can reproduce, from good music down to bad political speeches, and not after the event, but during it, and not in small doses, but at full length. The gramophone, however, has an advantage which appeals to many: you can listen to what you choose when you choose; not, as in wireless, to what someone else chooses when he chooses. There is no end to the possibilities of wireless broadcasting if we forget the limitations, and even with the limitations as they are now, there is a wide field. In the United States of America there are over a million receiving sets in use, and the demand continues.

There are also many thousand sending stations, which gave rise to the somewhat chaotic state of affairs already mentioned.

In America, music, stock exchange quotations, weather forecasts, health information, sermons, speeches and advertising matter are broadcasted, and when there was an absence of the control which is now being arranged, it is easy to imagine how frequently those at the receiving sets must have been served up with a remarkable bran pie.

In this small densely populated country the folly of letting loose any wireless broadcasting system without control is obvious. Here, all non-Government wireless stations for either sending or receiving have to be licensed by the Postmaster-General. Receiving licences are readily obtained, but transmitting licences, which are charged for according to the power used, outside those required by commercial wireless companies, are granted sparingly and under strict conditions to bona-fide experimenters only.

A licence to establish a wireless station for the purpose of receiving messages on wave-lengths between 300 and 500 metres, a range of waves which covers the broadcasting programmes will, it has been announced, be obtainable at a post office.

If a licence is required for the purpose of carrying out experimental work in either transmission or reception, application must be made to the Secretary of the General Post Office, London.

Facilities for broadcasting by wireless telephony have been granted to British wireless manufacturing firms. The matter broadcast may be instructive or purely entertaining, as songs and stories, and may be carried out from stations in the following centres: London, Manchester, Birmingham, Glasgow or Edinburgh, Aberdeen, Newcastle, Cardiff and Plymouth, at any time between 5 p.m. and 11 p.m. on weekdays, and at any time on Sundays, on wavelengths between 350 and 425 metres.

For reception, a suitable set can be purchased for from about £5 to £50, the better the set the better the results, and, like most other purchases, it is seldom economical or satisfactory to buy a cheap article if a more expensive one can be afforded.

It is better to have if possible an outside elevated wire connected to the receiver, but a self-contained, though more expensive, set placed inside the room can be quite as good as, or better than, a cheaper one with an outside wire.

This is a start. Where the future will land us it is impossible to say. We can all indulge in prophecies, but let them be at least intelligent; let us remember that jamming is caused by signals sent on the same wave-lengths or lengths close to it, or by electrical disturbances in the atmosphere, and that the waves radiate in all directions, but let us remember also that as the art advances jamming troubles will become less serious, and that directional working will become more practicable.

# CHAPTER X

## WIRELESS AND WIRES

THE general principles of line and of wireless working are very similar. In the former, electric currents are sent along a wire from the sender to the receiver, the energy being transmitted at a speed of about 186,000 miles a second, that is to say, instantaneously, for all practical purposes; and in the latter, electric waves are sent through space from the sender to the receiver at about the same speed.

The first great similarity between line and wireless is, therefore, the speed of travel: the first great dissimilarity, the connection used; in the one, a wire; in the other, space. The advantages of a wire are that the messages can be sent, if desired, to one receiver only, whereas in wireless the messages are sent into space, and can be picked up by receivers other than those for which they are intended. This fact, though obviously useful for transmitting simultaneously to many receivers, looks at first sight to be a very great drawback when secrecy is important, as is so often the case in the business of both Governments and of individuals. The progress of wireless in recent years has, however, improved matters to an extent not generally realized.

The chief development which assists materially towards secrecy is that of high-speed working. By high speed is meant a speed which is greater than can be sent by hand, and received by ear; in other words, high-speed working is such as entails the use of automatic transmitters and receivers. When working is carried out in this way the messages cannot be tapped without considerable difficulty, and the use of expensive apparatus. This gets rid of the casual eavesdropper, but does not prevent any well-equipped station within range from being able to read the

Engine room, Cairo Station.

[To face page 88.

messages, though, even then, the use of cipher instead of plain language will make it difficult for the intercepting station to obtain much information. Another development tending towards secrecy is directive transmission; that is to say, stations can now be designed to send messages in one direction better than in any other, so that the electric waves can be concentrated, to a certain extent, towards the receiving station instead of being radiated out into space in all directions. This bogey of secrecy as a disadvantage to wireless is, in fact, usually overdone, and especially is this apparent when comparing wireless with land-line working through countries of various nationalities, where the messages pass through many hands. In the case of direct submarine cable communication it is a different matter, and there is no doubt that from the secrecy point of view cables have a distinct advantage over wireless. High-speed wireless working has only now reached that stage of development when its use for everyday commercial purposes is becoming practicable, and at the present moment much of the long-distance wireless work is still being conducted at hand speed, though it is quite certain that, in a very short time, this will no longer be the case, as working at hand speed over long-distances is not a paying proposition. A number of long-distance wireless circuits are being worked, under good conditions, at about fifty words a minute, and some medium-range ones of a few hundred miles, at about one hundred words a minute. But this is only the commencement, and it is on this line of development that wireless may reasonably be expected to compare favourably with cables before very long. At present, the fifty words a minute under *good* conditions for wireless does not compare well with the fifty words a minute which can be obtained by long-distance cables under *normal* conditions, and the same may be said, and to a greater extent, of the hundred words a minute for wireless medium-range working, compared with land-line working.

Besides secrecy, however, the dissimilarity in the connection between sender and receiver, viz., wires and space, entails other important matters for comparison.

In the case of line working, obvious mechanical difficulties may occur, the greatest being that the wire may break, thus severing all communication for possibly a long period, unless there are alternative routes. Other troubles include, first, leaks of the current to "earth" due to electrical connections occurring between the wire and the ground or sea, and, secondly, the currents produced in the wire by atmospheric electrical discharges causing false signals, or, as it is called, "jamming" of the messages. Now, wireless is free from these troubles with the exception of jamming by "atmospherics," but this defect in the case of wireless is very serious indeed, and is, in fact, the greatest disadvantage of wireless compared with line working. Wireless messages are transmitted by creating electrical disturbances in space, and are thus easily "jammed" by natural electrical disturbances or discharges, such as occur frequently in warm weather in these latitudes, and very much more frequently and strongly in tropical countries. In extreme cases, as occasionally in some tropical countries, atmospherics are troublesome in line working, but in long-distance wireless they are always troublesome, except in temperate climates in the cool weather.

The number of infallible inventions for "eliminating" atmospherics are legion, but the trouble is still, unfortunately, sufficient to foster that suspicion under which long-distance wireless working has laboured ever since Senatore Marconi sent the first message across the Atlantic in December, 1901. There is no doubt that the jamming caused by atmospherics has been lessened to a very great extent during the last few years, and that progress in this direction should become even more rapid now that so many high-power stations have been completed and are available for experiments.

In the British Isles, at the moment, the only high-power stations are two transatlantic stations owned by the Marconi Company, viz., Carnarvon and Clifden, and one Post Office station at Leafield, near Oxford, which communicates, mainly, with the similar Post Office station near Cairo in Egypt. It is intended, however, as we have seen, to establish as soon as possible a chain of high-power stations for the Empire, in accordance with the recommendations made in 1920 by a Committee appointed by the Government, and the scheme is now being worked out in detail.

It must be remembered that this is really only the beginning of high-power wireless as a world-wide commercial proposition, and that much still remains to be done before long-range, or even medium-range, wireless services can give traffic returns up to the standard of cables or land lines, a standard which has only been obtained after many years of experience and experiments under actual working conditions. It is as easy as it is unprofitable to launch off into golden dreams of the future of long-distance wireless communications, dreams, which luckily for many of the dreamers, will have been long forgotten before the time can arrive when they will be put to the test. Many of us remember our own dreams when wireless messages were first sent across the Atlantic twenty-one years ago, and most of us will ask for another year or two, and then—well, perhaps it is better to "wait and see."

So far, we have been thinking of long-range telegraph working, but what of the telephone? Here we are on even more uncertain ground, because there are no long-range commercial wireless telephone stations at all in operation. Long-range land-line telephony is now, at last, coming to its own, due to the invention of the Thermionic Tube. In wireless, as we have seen, these tubes can be used for the transmission of signals or speech over long distances as well as for amplifying the current at the receiving station, that is, increasing the strength of the signals or speech received, in which latter manner they are used also in line telephony. Submarine telephony by cable is not practicable for these long ranges, and for transoceanic work and for the broadcasting of messages great developments in wireless telephony may be expected. At present, much use is being made of wireless telephony for communication with aircraft, and nearly all aircraft wireless sets can now be used either for telegraphic or telephonic purposes, as desired. Merchant ships are not yet being fitted to any extent with wireless telephony, and there is, in fact, no commercial service in operation anywhere over even short distances, with the exception of one between a lightship in the Mersey and the shore, and a few rather experimental services in the United States and on the Continent. The chief drawbacks to wireless telephony in its present stage of development are that the mutual interference, or jamming, of messages, as well as jamming by atmospheric electrical disturbances, are even greater than in the case of messages sent by wireless telegraphy, and that the consumption of electrical power is also greater. The first trouble means that jamming by atmospheric disturbances is very serious indeed, and that it is impracticable to have a large number of stations within range of each other in operation at the same time; the second, that the cost is higher than with wireless telegraphy. The Marconi Company have, however, carried out some very promising experiments, in co-operation with the Post Office, with wireless telephony to Holland on short wave-lengths by which the jamming difficulties are much reduced, an advance which augurs much for the future. The notices concerning wireless telephony which appear from time to time in the newspaper Press are invariably of the golden dream type, just as they were in the early days of wireless telegraphy; and as the business man finds that no facilities are offered to him for using a wireless telephone, so does he become suspicious as he did years before in the case of wireless telegraphy. He feels that something is wrong, and for want of another scapegoat he turns, as usual, to the everlasting stupidity of the Government of the day.

In looking at this somewhat sombre picture of the attainments of wireless telephony, we must not forget that it is only in its infancy and that it has two attributes of the greatest importance: first, the sender is connected to all receivers by space, which costs nothing; and, secondly, there is almost no voice distortion such as makes long-distance submarine telephony impracticable. With these outstanding fundamental advantages it is easy to foresee possibilities for wireless telephony far outdistancing those of line working for long-distance transoceanic communications, and for "broadcasting" messages.

For mobile communications, such as for ships and aircraft, there can be no comparison with line working, as wireless has a full-blown monopoly, a happy possession, though in this case one which certainly does not yet come within the magic circle of the wealthy industrial monopolies.

# Mullard

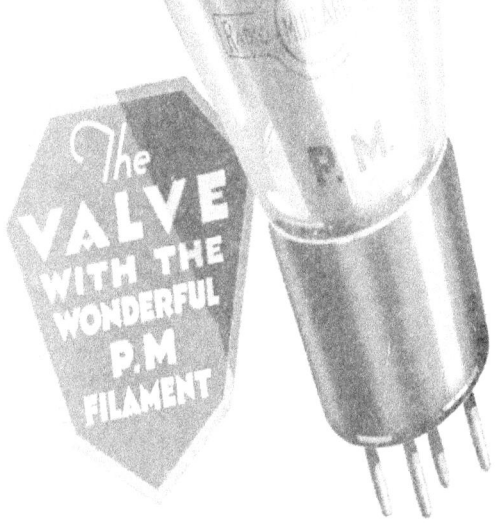

The VALVE WITH THE WONDERFUL P.M FILAMENT

P.M

THE · MASTER · VALVE